X-Ray Fluorescence Spectrometry (XRF) in Geoarchaeology

M. Steven Shackley

Editor

X-Ray Fluorescence Spectrometry (XRF) in Geoarchaeology

 Springer

Editor
M. Steven Shackley
Department of Anthropology
Geoarchaeological XRF Laboratory
University of California
232 Kroeber Hall
Berkeley, CA 94720-3710
USA
shackley@berkeley.edu

ISBN 978-1-4419-6885-2   e-ISBN 978-1-4419-6886-9
DOI 10.1007/978-1-4419-6886-9
Springer New York Dordrecht Heildelberg London

Library of Congress Control Number: 2010938434

Printed on acid-free paper

Springer is part of Springer Science+Business Media (www.springer.com)

"For Wilhelm K. Röntgen, Arthur Compton, Charles G. Barkla, Henry G. J. Moseley, Edward Hall, Ron Jenkins, Ian Carmichael, Joachim Hampel and all the other founders and facilitators of Archaeological XRF"

Preface and Acknowledgements

In many ways this volume has been a long time coming. While X-ray fluorescence spectrometry (XRF) has been in the literature for many years and the archaeological application since at least Edward Hall's (1960) paper in *Archaeometry*, we had not yet attempted to put it all together in a defined whole until now. X-ray fluorescence spectrometry in all its many forms – including the two focused on here, energy-dispersive X-ray fluorescence (EDXRF) and wavelength dispersive X-ray fluorescence (WXRF) – has been one of the most important technologies used by archaeologists to explain the past through many of its paradigmatic shifts from the Cultural Historical approach to the New or Processual Archaeology to Post-Processual Archaeology and to whatever normal science we are in now. Throughout these changes in the perspective on the past, archaeologists have increasingly relied on XRF as a tool that has been used to address so many of the problems of interpreting the past including, but certainly not limited to, lithic procurement, exchange, group interaction, social identity, gender relations, and many other areas. Through it all, XRF has been continually evolving from the older manual goniometer XRF instruments like I used in graduate school where the results of the peak heights were simply printed out on a teletype and one had to generate individual data reduction routines, to our sophisticated Windows-based software that leads us through elemental acquisition, standard library construction, calibration, and reliable results on instruments that are shrinking to hand-held sizes. As the computer and software revolutions have given us superior data analysis support, the hardware itself has improved. EDXRF detectors now can process the chaotic X-ray data through twenty-first century multi-channel analyzers, such that some of the elemental data are as precise as that acquired by neutron activation analysis (NAA) and other instrumentation in the geoarchaeological arsenal. It is now possible for me to teach undergraduates how to analyse samples in minutes with these advances in XRF technology. And yet, there is still an "art" of XRF that comes through in these chapters. Sure, the instrumentation software can lead you through an instrument set-up, but it does not show you how to place those pieces of obsidian angular debris on the sample tray, such that the largest amount of material is presented to the X-rays, or that zirconium numbers are better acquired when you can get the sample right down at the preferred point of irradiation.

As well as the art of XRF, the analyst must attempt to understand the archaeological problems as well. Gone are the days when physicists and chemists operated the instrumentation and provided the data to the archaeologist. Increasingly, as you can see in these chapters, archaeologists and geoarchaeologists are running these labs, and when physical scientists are doing the work, these are the scientists who have taken the time to become archaeological savvy. They attend archaeological meetings, even serving on executive committees of organizations like the International Association for Obsidian Studies (http://members.peak.org/~obsidian/index.html) and the Society for Archaeological Sciences (http://www.socarchsci.org/), not to mention participating in the International Symposium on Archaeometry (http://www.itarp.uiuc.edu/atam/newsandevents/intlarchaeometrysymposium.html) held every other year and in Tampa, Florida in 2010.

While this volume could have presented other instrumental technologies that are allied to XRF such as producing X-rays by synchrotron radiation, or the very good work with PIXE and PIXE-PIGME, we were more interested in presenting XRF in a more "pure" form and relating the ideas in a structure that could be understood by the users of XRF data, not the producers of XRF data. And actually, the basics of XRF can be extrapolated to any instrument that uses X-rays to produce fluorescence and measured through collimation, detectors, pulse processors, multi-channel analyzers, and sophisticated software.

We (actually I) decided to restrict the application of XRF to geological material, although Liritzis and Zacharias cover other materials as well in Chap. 6 on the comparison between XRF and PXRF. The vast majority of applications of XRF in archaeology are to geological materials, particularly obsidian and other volcanic rocks for very good reasons discussed in the various chapters. That was my decision and, perhaps, signals my bias in XRF studies, but I believe it is a reality in the archaeological world. One of the book reviewers made this point as well.

No undertaking like an edited volume uniting a number of scholars from around the world can be done by one person. This is certainly no exception. First, I must thank tremendously my chapter authors. I have put together edited volumes before that required infinite patience with authors – this was definitely not the case with these folks. Indeed, their drafts were in before I could even finish Chap. 2. And, I could not have written an introduction to XRF without these scholars. Some of the authors I have known for decades, some just recently, but all are well respected or rising stars in the discipline. A special thanks to Phil Johnson who worked under unusual circumstances to get his edited version back to me. Thank you all for making this volume real for archaeologists.

Springer's archaeology editor, Teresa Krauss is one of the most knowledgeable editors I have worked with. Also, Katherine Chabalko, who is probably the most dedicated archaeological science editor in the business and a pleasure to work with, is also editing sociology, so her knowledge of social science issues is remarkable.

My wife and scholar-partner Dr. Kathleen Butler is very much responsible for this volume getting done. She provides the time, space, and great listening to my complaining to the point of sainthood. I could not have done this, indeed any project, without your help and understanding.

The talented people at www.learnxrf.com were very helpful in giving advice, and I have used and modified their teaching Powerpoint for a number of years in my own pedagogy. Thank you.

I must thank my archaeological colleagues, especially my friends and colleagues in the American Southwest and Northwest Mexico who have made XRF work so much fun and rewarding through the years. Special thanks to my colleague Dr. Kent Ross in the Department of Earth and Planetary Sciences (now at NASA) for discussions about the vagaries of XRF and who commented on my introduction to XRF in Chap. 2. Tim Teague, also in EPS and one of the authors in Chap. 3, has been my constant colleague in the XRF realm and deserves much more credit than he generally receives. They all encourage my work and encourage me to do better work at all times.

Finally, my students in the XRF, Geoarchaeological Science classes, and archaeological petrology field schools not only keep my hope up that all this will continue evolving forever, but also give me so much to work for. I especially thank the Spring 2009 undergraduate and graduate students in the Archaeological XRF Lab method course who shaped much of what you read in Chap. 2. I am sorry for using you as my sounding board, and you deserve some of the credit here. Thanks especially to Esteban Gomez, who as the Graduate Student Instructor in that class, and the new Assistant Professor in Anthropology at Colorado College, for keeping it and me all together. Celeste Henrickson, the Graduate Student Instructor in my archaeological petrology field schools and my last XRF lab class at Berkeley, is one of the most effective teachers and geoarchaeologists I have known. After initial editing I learned that Mark Pollard (Oxford) reviewed this volume and made many excellent suggestions. Thanks as always Mark.

National Science Foundation grants in support of the laboratory (BCS 0716333) and the continuing support of the Southwest Archaeological Obsidian Project (DBS-9205506; BCS-0810448), the L.S.B. Leakey Foundation, the UCB Stahl Endowment for Archaeological Research, many universities, various Federal agencies in the United States, and INAH, Mexico, as well as Native American tribal organizations, and cultural resource management firms all provided support and impetus for this volume. I hope you like what we have done.

Albuquerque, NM, USA   M. Steven Shackley
Berkeley, CA, USA

Contents

Contributors

M. Bocci
Dipartimento di Scienze della Terra, Università della Calabria,
87036 Rende (CS), Italy

G.M. Crisci
Dipartimento di Scienze della Terra, Università della Calabria,
87036 Rende (CS), Italy

M. Kathleen Davis
Far Western Anthropological Research Group, 2727 Del Rio Place,
Suite A, Davis, CA 95618, USA

Annamaria De Francesco
Dipartimento di Scienze della Terra, Università della Calabria,
87036 Rende (CS), Italy

Arian Drake-Raue
Department of Anthropology, University of Hawaii-Hilo, Hilo,
HI 96720, USA

Michael D. Glascock
University of Missouri, Research Reactor Center, Columbia,
MO 65211, USA

Joachim H. Hampel(deceased)
Department of Earth and Planetary Sciences, University of California,
307 McCone Hall, Berkeley, CA 94720-4767, USA

Thomas L. Jackson
Pacific Legacy, Inc, 1525 Seabright Avenue, Santa Cruz, CA 95062, USA

Phillip R. Johnson
Department of Anthropology, Texas A&M University, College Station,
TX 77843, USA

Rosemary A. Joyce
Department of Anthropology, University of California,
232 Kroeber Hall, Berkeley, 94720-3710, USA

Scott Kekuewa Kikiloi
Department of Anthropology, University of Hawaii-Manoa,
2500 Campus Road, Honolulu, HI 96822, USA

Ioannis Liritzis
Department of Mediterranean Studies, University of the Aegean,
Laboratory of Archaeometry, Rhodes 85100, Greece

Steven P. Lundblad
Department of Geology, University of Hawaii-Hilo, Hilo, HI 96720, USA

Peter R. Mills
Department of Anthropology, University of Hawaii-Hilo, Hilo,
HI 96720, USA

M. Steven Shackley
Geoarchaeological XRF Laboratory, Department of Anthropology,
University of California, 232 Kroeber Hall, Berkeley 94720-3710, USA

Timothy Teague
Department of Earth and Planetary Sciences, University of California,
307 McCone Hall, Berkeley, CA 94720-4767, USA

Nikolaos Zacharias
Department of History, Archaeology and Cultural Resources Management
Laboratory of Archaeometry, University of the Peloponnese,
24 100 Kalamata, Greece

Chapter 1
X-Ray Fluorescence Spectrometry in Twenty-First Century Archaeology

M. Steven Shackley

X-ray fluorescence is now a well-established method of analysis both in the laboratory and industry. The fact that the method is essentially non-destructive makes it particularly attractive for the analysis of archaeological and museum artifacts. Due to certain fundamental characteristics of the technique it is not suitable for some projects which would seem at first sight to present no problems (Hall 1960).

Edward Hall's abstract for his 1960 paper entitled "X-ray fluorescent analysis applied to archaeology" in the journal *Archaeometry* is just as appropriate half a century later. X-ray fluorescence spectrometry (XRF) is even more "well established" now, but is "not suitable for some projects" even though it might seem so, and archaeologists might think XRF is really appropriate. This volume is dedicated to issues in XRF analysis in geoarchaeology in particular. How does XRF work, and more importantly when and where is it appropriate? We have attempted to convey this without using physical science jargon, although it was difficult at many points. I have provided a glossary at the end of the volume to help in this direction.

Today the market is being flooded with, it seems, hundreds of portable X-ray fluorescent instruments (PXRF), but do they really do all that the marketing suggests? The recent edited volume by Potts and West (2008) is devoted to the portable instrument for a variety of applications in science and engineering, including archaeological stone and museum works of art (Cesareo et al. 2008; Williams-Thorpe 2008), and our own Ioannis Liritzis and Nikolaos Zacharias have provided a critical evaluation of PXRF in obsidian studies in this volume (Chap. 6). Is there a need anymore for the desktop laboratory EDXRF? Can we as archaeologists get away with only a much less expensive PXRF? As an academic in a major university, I see the upcoming students' increasing interest in PXRF as a solution to a problem – is it a solution for a problem, or a very real need to be able to carry great

M.S. Shackley (✉)
Department of Anthropology, University of California, 232 Kroeber Hall, Berkeley, CA 94720-3710, USA
e-mail: shackley@berkeley.edu

M.S. Shackley (ed.), *X-Ray Fluorescence Spectrometry (XRF) in Geoarchaeology*, DOI 10.1007/978-1-4419-6886-9_1, © Springer Science+Business Media, LLC 2011 analytical power into remote field conditions? It may, indeed, be too soon to definitively answer these questions, but we will try.

I think we have assembled a group of chapter authors who exemplify the very real ideas and issues that XRF has been grappling with for many years, both in XRF theory and method, and for a variety of archaeological materials. And yes, this volume is focused on XRF applied to geological materials, and not ceramics or organic materials in archaeology. The vast majority of archaeological XRF world-wide is applied to these inorganic materials, and there are very good technical reasons for that covered in the next chapter and the Davis et al.'s chapter on EDXRF method (Chap. 3). This does not mean, of course, that there are no perfectly good reasons for analysing ceramics or organics. While pottery is "not a rock," it is composed of rocks and clay sediments that lend themselves to XRF analysis (Hall 2001; Neff 1992; Pollard 1996). The very real difference is that unlike volcanic rocks as an example, they are not formed at very high temperatures in the mantle or crust and are not necessarily a reflection of production. The sometimes perverse human behaviour that creates pottery is not an issue with a material produced at more than 1,000°C by nature. With many organics, particularly liquids, a helium flush is necessary in the X-ray chamber in order to analyse for the light elements and eliminate contamination from the atmosphere. Nevertheless, the concepts related in this volume are generally, sometimes precisely, the same as ceramic and organic analyses.

What is very different from Edward Hall's XRF world is that, archaeology as a discipline has changed markedly both theoretically and methodologically, due in large part to the tools and technology that are offered in the twenty-first century. Issues of social identity, equality, gender, national character, and native rights are all current and important issues in archaeology today, and as we can see, are issues that are more efficiently addressed with these new tools – XRF certainly among them. I have argued elsewhere that without our twenty-first century archaeological tools, a twenty-first century archaeology would not be much more than the New Archaeology of the 1960s or Culture Historical in nature (Shackley 2005, 2008). XRF will play an important role as a "well established method" for illuminating our ever-changing problems in archaeology well into this century. What XRF will look like in another 50 years is impossible to imagine. The portables are probably a peep into that future.

What Is in this Volume?

The authors that I chose for this volume were not chosen randomly. They are colleagues and in some cases longtime friends of course, but they run the gamut from those who have been working in the discipline for decades, and those younger, bright new faces in the XRF world. I believe that a discipline, even emerging ones, benefits by the wisdom and experience of the old cantankerous scholars like me, tempered by the energy and new ideas of the younger scholars. And while this seems upside down in typical psychology, my experience of over 20 years in an academic department suggests to me that the tempering in a field is by the young and not the old towards the young. Nevertheless, I think we have organized ourselves in this volume in a readable and interesting way for those archaeologists, geoarchaeologists, and earth scientists who are curious about XRF in archaeology and where it might be going in a twenty-first century science.

First, I endeavour to explain the physics, mechanics, and method of XRF for educated and curious archaeologists and others in Chap. 2. This was a chapter I have wanted to write for sometime, but my academic coursework in XRF took 25 years and my notes were extremely out of date (our text was Jenkins 1981 book). I also attempted to look beyond the "normal science" of XRF I have done for the last quarter century and incorporate the newer instrumentation and methods.

How does XRF work? This is, of course, the primary point of Chap. 2, but I have used a rather linear examination of the subject, from a short history, to the basic physics of X-rays, to spectral examination, types of instrumentation, interference effects, calibration theory, and finally a word about presentation of data and the instrument method used at the Berkeley XRF lab. Luckily for me, I teach an undergraduate method course in XRF and was able to bounce these ideas off anthropological archaeology students, most with little background in physical science.

Chapter 3 is basically a reprint of the Davis et al.'s (1998) paper that in many ways was the germ for this volume. Kathy Davis was a brilliant young student of XRF who designed experiments with size and surface configuration of obsidian samples that form the basis of the "normal science" of EDXRF technique today. Aspects of the 1998 work can be found in all the chapters in this book. What is the smallest size that can be analysed non-destructively by laboratory EDXRF? More recent works by Lundblad et al. (2008), and Eerkens et al. (2007) have grappled with the issue, in the first case for basalt artifacts in the Pacific Basin, and in the second case for hunter-gatherer artefact assemblages in the American Great Basin. Importantly, the conclusions of Davis et al. in 1998 have stood the test of time, and for this reason it is presented again with only slight modifications.

While EDXRF has been used mainly for obsidian studies in archaeology, a number of researchers have been just as interested in basalt artifacts, particularly in Oceania, where basalt artifacts were transported throughout the Pacific Basin in prehistory. In Chap. 4, Steve Lundblad, Peter Mills, Arian Drake-Raue, and Scott Kekuewa Kikiloi of the EDXRF lab at the University of Hawaii-Hilo detail their method for non-destructive EDXRF analyses of this important stone in this huge region. In a former paper, as mentioned above, some of these scholars experimented with the size and surface issues in basalt as Davis et al. had 10 years earlier with obsidian (Davis et al. 1998; Lundblad et al. 2008). Surface weathering is a concern that these authors must deal with in tropical environments as they do admirably. The understanding of the social ramifications of the exchange of basalt axes in this region is well served by this work.

While the majority of chapters in the volume are devoted to some aspect of EDXRF, Anna Maria De Francesco, M. Bocci, and G.M. Crisci look at the potential for non-destructive wavelength XRF of archaeological obsidian in Chap. 5. They re-visit the possibility of using semi-quantitative (elemental ratios) results in the WXRF analysis of obsidian and conclude that, in some instances in the Mediterranean, it is possible to confidently assign obsidian to source with semi-quantitative data. As I discuss in Chap. 2, semi-quantitative WXRF results were tried early on in archaeology, but were abandoned when quantitative results were possible with EDXRF and WXRF by rationing to the Compton scatter, and the software and algorithms were available to do that. In Chap. 5, the authors resurrect semi-quantitative analyses and show how it can work in some regions.

Portable EDXRF (PXRF), as mentioned above, is revolutionizing source provenance work in archaeology, or is it? Ionnis Liritzis and Nikolas Zacharias discussed a comparison between PXRF and desktop XRF results on the same material in an effort to deal with this in a controlled way in Chap. 6. Craig et al. performed a direct experiment comparing the results of PXRF and desktop XRF in a 2005 paper for Andean obsidian with some interesting results, but many questions (see also Phillips and Speakman 2009 for a more recent study in Russia). In Chap. 6, the authors look at PXRF applications worldwide for a wide variety of materials including natural and man-made materials such as ceramic, glaze, glass, obsidian, pigments, paint, and metal artifacts, as well as a direct comparison with their own data. They conclude, as many have today, that PXRF, while not providing the sensitivity that may be needed for many applications, has a role to play in archaeology today, and the quality of PXRF data will certainly improve in the future.

For most of my career, XRF has been compared to neutron activation analysis (NAA) favourably and unfavourably in the literature, at times with experimentation and at times as merely opinion (Hughes 1984; Shackley 1998, 2005). We accept that for non-destructive analyses, XRF is one of the few analytical alternatives. For the greatest instrumental precision and accuracy of results, NAA is certainly the best course to take. But does EDXRF really provide data that are equal to NAA in providing results that are defensible analytically and archaeologically? The next two chapters go a long way in answering this question. These two chapters are direct comparisons between these two powerful tools on the same artifacts – basalt and obsidian.

As noted in the discussion of Chap. 4, the XRF analysis of basalt is becoming much more common in archaeological applications. In Chap. 7, in order to see whether XRF is really comparable to NAA in the analysis of basalt artifacts from Samoa, Philip Johnson presents an important direct comparison between the results of both EDXRF and NAA on the same basalt artifacts. Can XRF of basalt provide the same reliable and valid data that the NAA surely can? His conclusion through this experimentation is that, in nearly all cases, XRF can provide useful data, particularly when the artifacts cannot be destroyed. This, as with the Lundblad et al. study in Chap. 4, supports the role of EDXRF analyses of basalt in addressing the issues of exchange and social interaction in the Pacific Basin.

Michael Glascock, one of archaeology's greatest physical scientists and one of the NAA experts in the field, provides a definitive direct comparison between NAA and EDXRF of obsidian sources in Central America (Chap. 8). Besides the publishing of extremely important data on these sources, his experiment taken with Phil Johnson's in the preceding chapter provides what I consider the defining characterization of EDXRF vs. NAA. Perhaps one of the most useful tabular presentations in the volume is Glascock's Table 8.1, a tabulated comparison between the relative merits of both technologies. What is most important for the reader to understand is that, this is just not an either/or issue. Glascock makes it clear that there are many regions, Central America as a good case, where the elemental concentrations that are available through XRF may allow a scholar to confidently separate some sources. Then, it would be necessary, if that information is crucial to addressing an archaeological problem, that NAA would be required to address some sub-source issues (see also Glascock et al. 1998). In a recent study that involved the Missouri Research Reactor Center, Eerkens et al. (2007) also discovered that samples smaller than the <10 mm threshold XRF were previously deficient; the analysis of obsidian debitage smaller than 10 mm indicated both greater diversity of sources and greater distances to source in those hunter-gatherer contexts in the American Great Basin. Neutron activation is not hampered by small sample size issues and could provide data with modern collection techniques in archaeology (water screening/flotation) that could change our concept of exchange, group interaction, and procurement like never before. The size limitation with EDXRF is not as great as it was 20 years ago. With tube collimation, we are getting good results down to about 2 mm sample sizes (see Chap. 2).

Glascock finally concludes that one can have "high confidence that data on obsidian samples measured by XRF can be compared directly to NAA collected by NAA". No study, until this time, provides this kind of result with such in-depth experimentation.

In the concluding Chap. 9, Rosemary Joyce, a renowned Mesoamerican archaeologist who is interested not only in the science of archaeology, but also in the social possibilities of the past and archaeology as a discipline, provides a look at the role of XRF in archaeology from the viewpoint of an archaeologist. Rosemary Joyce has looked at and published extensively on all the issues I mentioned as current in archaeology today – social identity, equality, gender, national character, and native rights. In many ways Rosemary's chapter is the purpose of this volume; not just to elucidate the issues in XRF in archaeology, but to provide the basic understanding for all archaeologists so that a real integration of archaeological materials science and social archaeology becomes the "normative" view.

The authors in this volume are dedicated XRF analysts, and we certainly have our biases, although we have endeavoured to make those biases as transparent as possible. However, we can sometimes forget that our data are used for a higher purpose by the archaeological community; indeed by the community that funds and simply makes our laboratories possible. We hope that this volume will be of value to any archaeologist with an interest in compositional analysis.

References

Cesareo, R., Ridolfi, S., Marabelli, M., Castellano, A., Buccolieri, G., Donativi, M., Gigante, G.E., Brunetti, A., and Medina M.A.R., (2008), Portable systems for energy-dispersive X-ray fluorescence analysis of works of art. In Potts, P.J., and West, M. (Eds.), Portable X-ray fluorescence spectrometry: capabilities for in situ analysis, (pp. 206-246). Cambridge: The Royal Society of Chemistry.

Davis, M.K., Jackson, T.L., Shackley, M.S., Teague, T., and Hampel, J.H., (1998), Factors affecting the energy dispersive X-ray fluorescence (EDXRF) analysis of archaeological obsidian. In Shackley, M.S. (Ed.), Archaeological obsidian studies: method and theory, (pp. 59–80). New York: Plenum.

Eerkens, J.W., Ferguson, J.R., Glascock, M.D., Skinner, C.E., and Waechter, S.A., (2007), Reduction strategies and geochemical characterization of lithic assemblages: a comparison of three case studies from Western North America. *American Antiquity*, 72, 585–597.

Glascock, M.D., Braswell, G.E., and Cobean, R.H., (1998), A systematic approach to obsidian source provenance. In Shackley, M.S. (Ed.), Archaeological obsidian studies: method and theory, (pp. 15–66). New York: Plenum.

Hall, E.T., (1960), X-ray fluorescent analysis applied to archaeology. *Archaeometry* 3, 29–37.

Hall, M.E., (2001), Pottery styles during the early Jomon Period: geochemical perspectives on the Moroiso and Ukishima Pottery Styles. *Archaeometry* 43, 59–75.

Hughes, R.E., (1984), Obsidian studies in the Great Basin: problems and prospects. In Hughes, R.E. (Ed.), Obsidian studies in the Great Basin, (pp. 1–26). Berkeley: Contributions of the Univesity of California Archaeological Research Facility 45.

Jenkins, R., (1981), X-ray fluorescence spectrometry. New York: Wiley.

Lundblad, S.P., Mills, P.R., and Hon, K., (2008), Analysing archaeological basalt using non-destructive energy-dispersive X-ray fluorescence (EDXRF): effects of post-depositional chemical weathering and sample size on analytical precision. *Archaeometry,* 50, 1–11.

Neff, H. (Ed.), (1992), Chemical characterization of ceramic pastes in archaeology. Monographs in World Archaeology 7. Madison: Prehistory Press.

Phillips, S.C., and Speakman, R.J., (2009), Initial source evaluation of archaeological obsidian from the Kuril Islands of the Russian Far East using portable XRF. *Journal of Archaeological Science* 36, 1256–1263.

Pollard, A.M., (1996), The geochemistry of clays and provenance of ceramics. In Pollard, A.M., and Heron, C. (Eds.), Archaeological chemistry, (pp. 104–147). Cambridge: Royal Society of Chemistry.

Potts, P.J., and West, M. (Eds.), (2008), Portable X-ray fluorescence spectrometry: capabilities for in situ analysis. Cambridge: The Royal Society of Chemistry.

Shackley, M.S., (1998), Current issues and future directions in archaeological volcanic glass studies: an introduction. In Shackley, M.S. (Ed.), Archaeological obsidian studies: method and theory. (pp. 1–14). New York: Plenum.

Shackley, M.S., (2005), Obsidian: geology and archaeology in the North American Southwest. Tucson: University of Arizona Press.

Shackley, M.S., (2008), Archaeological petrology and the archaeometry of lithic materials. *Archaeometry,* 50, 194–215.

Williams-Thorpe, O., (2008), The application of portable X-ray fluorescence analysis to archaeological lithic provenancing. In Potts, P.J., and West, M. (Eds.), Portable X-ray fluorescence spectrometry: capabilities for in situ analysis, (pp. 174–205). Cambridge: The Royal Society of Chemistry.

Chapter 2
An Introduction to X-Ray Fluorescence (XRF) Analysis in Archaeology

M. Steven Shackley

As I have discussed in the last chapter, our goal here is not to elucidate XRF for the entire scientific community – this has been done admirably by others – but to translate the physics, mechanics, and art of XRF for those in archaeology and geoarchaeology who use it as one of the many tools to explain the human past in twenty-first century archaeology. While not a simple exercise, it has utility not only for those like us, who have struggled (and enjoyed) the vagaries of XRF applications to archaeological problems, but for a greater archaeology. First, we trace the basic history of X-rays used in science and the development of XRF for geological and archaeological applications, and the role some major research institutions have played in the science. Following this is an explanation of XRF that, in concert with the glossary, illuminates the technology.

History of XRF in Archaeology

X-rays were first discovered by the German physicist Wilhelm K. Röntgen (1845–1923) for which he won the Nobel Prize in 1901 (Röntgen, 1898). While X-rays have been used for commercial elemental analysis since the 1950s, X-ray spectroscopy is much older than that, dating back to 1909 when Charles G. Barkla found a connection between X-rays radiating from a sample and the atomic weight of the sample. In 1913, Henry G. J. Moseley helped number the elements with the use of X-rays, by observing that the K line transitions in an X-ray spectrum moved the same amount each time the atomic number increased by one (Moseley, 1913/14). He is credited with the revision of the periodic tables, which were based on increasing atomic weight, to periodic tables based on atomic number. He later laid the foundation

M.S. Shackley (✉)
Department of Anthropology, University of California, 232 Kroeber Hall, Berkeley, CA 94720-3710, USA
e-mail: shackley@berkeley.edu

M.S. Shackley (ed.), *X-Ray Fluorescence Spectrometry (XRF) in Geoarchaeology*, DOI 10.1007/978-1-4419-6886-9_2, © Springer Science+Business Media, LLC 2011 for identifying elements in X-ray spectroscopy by establishing a relationship between frequency (energy) and the atomic number, a basis of X-ray spectrometry.

The potential of the technique was quickly realized, with half of the Nobel Prizes in physics awarded to developments in X-rays from 1914 to 1924. Originally, X-ray spectroscopy used electrons as an excitation source, but the requirements of high vacuum, electrically conducting specimens, and the problem of sample volatility posed major roadblocks. To overcome these problems, an X-ray source with a metal target was used to induce the fluorescent emission of secondary X-rays in the sample. Excitation of the sample by this method introduced some problems by lowering the efficiency of photon excitation and requiring instrumentation with complex detection components. Despite these disadvantages, the fluorescent emission of X-rays would provide the most widely used tool for the analyst using commercial instruments.

Why Non-Destructive X-Ray Fluorescence Spectrometry?

XRF hardware, the design of this instrumentation, and the decisions made in the selection of a particular instrument are discussed later. The overarching assumption in this volume is that XRF, particularly energy-dispersive X-ray fluorescence (EDXRF) spectrometry, solves many of our problems in geoarchaeology. Mike Glascock covers this in some detail in Chap. 8, in a comparison of EDXRF with NAA. Here, the real positive and negative points of XRF in archaeology are discussed.

What is Good About XRF?

The appeal of X-ray analysis of archaeological specimens lies in its remarkable combination of practical and economic advantages:

- Non-destructive
  In the vast majority of cases, analysed samples are not destroyed or changed by exposure to X-rays. They can thus be saved for future reference or used for other types of testing that may be destructive, such as obsidian hydration analysis.
- Minimal preparation
  Many samples can be examined with little or no pre-treatment, including almost all obsidian artifacts. Many of the alternative techniques require dissolution procedures that are both time-consuming and costly in terms of the acids or other reagents required. While it is best to wash any sediments off archaeological specimens it has been shown that if the dirt is minimal, and the artefact has not been subjected to heat so high as to melt some sediment matrix onto the sample, vigorous cleaning is not necessary (Shackley and Dillian, 2002). This is mainly due to the penetration of X-rays in the mid-Z X-ray region beyond the surface,

and while it does incorporate any contamination on the surface it is generally not an issue if some soil remains in the flake scars. The analysed volume is very large compared to any surface contamination. This is not the case with most metals, where patination and chemical weathering can radically change the composition at the surface and yield erroneous results (Hall, 1960).

- Fast
  X-ray spectrometry enables chemical compositions to be determined in seconds. For an analysis of the elements Ti-Nb on the Berkeley Spectrace and Thermo desktop instruments, at 200 live seconds per sample it takes about 5–6 min per sample depending on mass.
- Easy to use
  Modern instruments run under computer control, with effective software to handle measurement set-up and results calculation. Tasks that once required the constant attention of a trained analyst can now be handled by skilled students and are fully automated (cf. Rindby, 1989; Lachance and Claisse, 1994).
- Cost-effective
  Without the more involved sample preparation necessary in most WXRF and all destructive analyses, the cost is significantly lowered per sample.

While this suggests that XRF will solve all our problems, it is not the all-knowing black box we would like it to be (see Bouey, 1991).

What Non-Destructive EDXRF Will Not Do

- Sample size limits: Samples >10 mm in smallest dimension and >2-mm thick are optimal for EDXRF analyses (see Davis et al., 1998; Lundblad et al., 2008; Chap. 3 here). Why is this important? Shackley (1990) and, more recently, Eerkens et al. (2007) noted that for hunter-gatherers in the North American West, high residential mobility often requires that stone sources, including obsidian, be conserved for long periods of time. As an example, an archaic hunter will attempt to rejuvenate a dart point rather than making a new one whenever possible as he or she moves through the landscape. The rejuvenation of that point creates debitage that is quite small, often smaller than 10 mm. With modern recovery techniques, these small debitage are recovered much more often than they were in the past, and to make a long story short, Eerkens et al. (2007) found that, indeed, these small obsidian flakes in Great Basin sites indicated not only a greater distance from the original tool raw material, but also a greater diversity of sources than would be visible in an analysis of the larger flakes with EDXRF. So, while EDXRF can analyse much of the stone material left in prehistory, it may not solve all problems of interest to twenty-first century archaeologists. To be fair, however, the procurement ranges that could be reconstructed were relatively accurate with samples above the 10-mm threshold in the Eerkens et al. (2007) study, and NAA, which is essentially a destructive technique (see Chap. 8), had to be used on the smaller flakes.

Recently, however, thanks to the newer digital EDXRF instrumentation and tube collimation, we have been able to reach down to sizes of 2 mm with the ThermoScientific Quant'X EDXRF at Berkeley (see also Hughes, 2010).

- Restricted elemental acquisition: As discussed below, non-destructive XRF is restricted generally to a subset of the mid-Z X-ray region, the best portion including Ti-Nb, contains excellent incompatible elements for volcanic rocks (Cann, 1983; Shackley, 2005). While some rare earth elements and those with low atomic numbers or with very low concentrations can be useful in discriminating sources, in most cases XRF cannot solve that problem. This is discussed in detail in Glascock's comparison between XRF and NAA in Chap. 8.
- XRF cannot characterize small components – XRF like NAA is a mass analysis – every component in the irradiated substance is included in the analysis. It is possible to collimate the incoming X-rays from the tube and/or into the detector to focus on small components such as various minerals, but environmental scanning electron microscopy (ESEM), electron microprobe, or laser ablated-inductively coupled plasma-mass spectrometry (LA-ICP-MS) is much better suited for this kind of analysis. However, the bulk analysis of volcanic rocks has been shown to be quite effective as we argue in this volume.

It is true that XRF will not solve all our problems in archaeological provenance studies; it is simply the best non-destructive analytical tool at our disposal at this time.

Commercial X-Ray Spectrometry

Early on three types of spectrometers were available to the analyst. From the 1950s to 1960s nearly all the X-ray spectrometers were wavelength dispersive spectrometers, such as those used initially at Berkeley, and by Shackley at Arizona State University (Jack and Carmichael, 1969; Hall, 1960; Hughes, 1984; Jack and Heizer, 1968; Shackley, 1988, 1990). In a wavelength dispersive spectrometer, a selected crystal separates the wavelengths of the fluorescence from the sample by diffraction, similar to grating spectrometers for visible light. The other X-ray spectrometer available at that time was the electron microprobe, which uses a focused electron beam to excite X-rays in a solid sample as small as 10^{-12} cm^3. The first microprobe was built by R. Castaing in 1951 and became commercially available in 1958. By the early 1970s, energy dispersive spectrometers became available, which use Li-drifted silicon or germanium detectors. The advantage these instruments brought to the field was the ability to measure the entire spectrum simultaneously. With the help of computers, deconvolution methods can be performed to extract the net intensities of individual X-rays more on that later.

Early Berkeley XRF Studies

One of the earliest, if not the earliest application of XRF, particularly EDXRF, in archaeology was at Berkeley; it was based on the primary WXRF work of Edward Hall at Oxford (1960). In 1960, Hall reported using a wavelength XRF at Oxford on Imperial Roman coinage, noting the problems with patination on these coins. Hall's paper still serves as a model for the issues surrounding XRF analyses of archaeological material. However, as Shackley has noted elsewhere, the University of California, Berkeley has played a major role in the application of XRF to archaeological problems (Shackley, 2005).

While it might seem egocentric of the editor to favour Berkeley's role in XRF analyses in archaeology, it is true that the Departments of Earth and Planetary Sciences (formerly Geology and Geophysics) and Anthropology have continually utilized XRF and particularly EDXRF for archaeological applications since the late 1960s. In the 1970s and again from the 1990s into the twenty-first century, over 90% of the XRF applications were and are in archaeology and geoarchaeology at Berkeley. There are, of course, very real technological and paradigmatic reasons for this. Geological and petrological theory in this new century are increasingly demanding greater and greater precision in analyses, a precision that XRF cannot offer compared with NAA or ICP-MS (see Chap. 8). Due to exciting new concepts in the relationships between mantle and crustal geology, isotope chemistry has increasingly supplanted much elemental chemistry in geology (see Weisler and Woodhead, 1995 for an archaeological example). Indeed, it seems that while geologists are still "XRF users", archaeologists are the XRF cadre in science these days. So, as at Berkeley, XRF in many institutions has become the purview of geoarchaeological science. Many advances in analytical XRF are coming from archaeology and geoarchaeology. The computer industry, engineering, construction (concrete) and the aircraft industry are still heavy users. Still, archaeology is one of the major "buyers" at Thermo Scientific for the Niton portable XRF (PXRF) instruments, and the QuanX, now Quant'X desktop instruments as well as very many other manufacturer's instruments. Three of the four EDXRF instruments that were used for studies in this volume are ThermoScientific (then ThermoNoran) QuanX or Quant'X machines or the earlier Spectrace instruments (purchased by Thermo). These laboratories are, in part, daughter labs of the 1970s Spectrace 440 instrument at Berkeley, and the original DOS software which has now been re-written as the WinTrace™ Windows application. As discussed below, many Masters and Ph.D. studies in archaeology, as well as Shackley's 1990s work at Berkeley, began on this Spectrace 440 EDXRF instrument. The first experiments on size and surface constraints in obsidian artefact studies with EDXRF were performed on this instrument and they formed the basis for Lundblad et al.'s Chap. 4 (Davis et al., 1998; Hampel, 1984; Jackson and Hampel, 1992; Shackley and Hampel, 1992; Chap. 3). Lundblad et al.'s Chap. 4 is based on the digital ThermoScientific QuanX instrument that was mainly championed at the University of Hawaii-Hilo by Professor Peter Mills, a Berkeley Ph.D. in anthropology. This is not an attempt to support this particular EDXRF instrument, but to highlight the interconnectedness of EDXRF in archaeology and the primary role that Berkeley has played in its dominance in geoarchaeology today.

In 1968, Robert Jack and Robert F. Heizer (Departments of Geology and Geophysics, and Anthropology, UC, Berkeley) published the first X-ray fluorescence (XRF) spectrometric analysis of archaeological obsidian in the New World. The next year (1969), Jack and Ian Carmichael (Department of Geology and Geophysics, UC, Berkeley, now the Department of Earth and Planetary Science – EPS) published "The Chemical 'Fingerprinting' of Acid Volcanic Rocks". And while Cann and Renfrew (1964) had 4 years earlier published their NAA characterization of Mediterranean obsidian, Berkeley's XRF analysis of obsidian artifacts for source provenance was the first in the New World, and the first of a multitude of XRF obsidian projects at Berkeley. For over 40 years now, Berkeley has remained a centre for obsidian studies using XRF spectrometry (Jack, 1971, 1976; Jack and Carmichael, 1969; Jack and Heizer, 1968; see also Asaro and Adan-Bayewitz, 2007; Giaque et al., 1993 for the XRF work up the hill at Lawrence Berkeley National Lab). Robert Jack analysed over 1,500 obsidian artifacts worldwide during this period with Jackson (1974), and then Richard Hughes (1983, 1984), who began to focus on California and Great Basin studies. The list of ceramic, obsidian, and other rock provenance studies since that time in which Berkeley XRF facilities were used by faculty, graduate students, undergraduate students and scholars from other universities would fill pages (see Hughes 1983, 1984; Jackson 1974, 1986; Shackley, 2005). Since 1990, the Geoarchaeological XRF Laboratory at UC, Berkeley, has analysed tens of thousands of artifacts, mostly obsidian and other volcanic rocks, supporting faculty, student, government and cultural resource management studies worldwide, and particularly from the North American Southwest (see Shackley, 2005). While the early studies were primarily focused on developing source standard databases for various regions of the world to permit the identification of stone tool raw material sources, more recent graduate student and senior scholar research that uses these facilities is now integrating obsidian provenance studies into current archaeological theory and method in western North America, South America, East Africa, Oceania and Mesoamerica (Dillian, 2002; Hull, 2002; Joyce et al., 2004; Kahn, 2005; Negash and Shackley, 2006; Negash et al., 2006; Shackley, 1991, 1992, 1995, 1998a, 1998b, 2005; Silliman, 2000; Weisler, 1993). As we discuss in this volume, while other techniques have been shown to exhibit more instrumental precision, XRF, particularly energy-dispersive XRF, has remained the leader in non-destructive studies of artifacts (see Davis et al., 1998; Hughes, 1983; Shackley, 1998c, 2005; cf. Speakman and Neff, 2005).

The Portable EDXRF Revolution

As mentioned in the previous chapter, PXRF instrumentation is beginning to transform archaeological science. Indeed, many disciplines that need rapid in-field or museum compositional analyses are looking at PXRF. In 2008, Potts and West published the edited volume "Portable X-ray Fluorescence Spectrometry" in the Royal Society of Chemistry series. And while the vast majority of chapters are devoted to physical science applications, two are focused on archaeological and museum applications (Cesareo et al., 2008; Williams-Thorpe, 2008). Recently there have been a number of comparative studies between lab/desktop EDXRF and PXRF with varying results (Pessanha et al., 2009). Craig et al. (2007) analysed the same prehistoric Andean obsidian artifacts through a partial blind test using the Berkeley *QuanX* EDXRF and a PXRF at MURR at the University of Missouri with what the authors considered good agreement in the mid-Z X-ray region (see discussion of X-ray regions below; Craig et al., 2007). However, two issues arose with the above study: (1) While there was a general, statistically significant agreement between the studies overall, significant differences occurred between EDXRF and PXRF in certain mid-Z elements; and (2) the error rate was noticeably higher, giving larger dispersions about the mean in biplots with PXRF (Craig et al., 2007). This was explained in part by differing calibration routines. However, my experience with the Thermo/Niton PXRF at Berkeley indicates that the calibration routine is quite similar to the ThermoScientific calibration routine in the WinTrace™ software for the Quant'X desktop, although the software is completely different in the Niton (we use 11 standards for calibration in the Niton, and used 13 for the older *QuanX* in the Craig et al., 2007 study).

Recently Pessanha et al. (2009) directly compared the matrix effects between PXRF and laboratory WXRF and found that the behaviour between the instruments was similar, although the PXRF presented a "tremendously high background when compared to the stationary [WXRF] one ... and some of the trace elements were almost not detected" (2009:497). This is a phenomenon we have found in the Berkeley NITON and the loaned Bruker instrument.

It is important for archaeologists to be aware that currently marketed PXRF instruments are not empirically calibrated out of the box, like desktop instruments (see calibration discussion below). Until quite recently, both off the shelf Bruker and Niton systems were calibrated through a fundamental parameters routine or not at all, which is fine for presence/absence analyses, but not adequate for the level of accuracy needed in most geoarchaeological studies. I realize how tempting it is for archaeologists to purchase a relatively inexpensive PXRF instrument that will, seemingly, solve all of their problems. I have had two former students purchase or borrow two different PXRF instruments; both were told that the "instruments are calibrated", and the results were disastrous. Just like any desktop instrument, in order to establish reliability of results, you must create an empirical calibration, with appropriate conditions and using international standards. Otherwise, the results may be internally consistent, as seen in the Craig et al. (2007) study, but could be incomparable to other studies that are or are not empirically calibrated. The new PXRF industry is definitely based on a *caveat emptor* philosophy. There indeed, may be a PXRF revolution in geoarchaeology, but it requires the same instrument set-up procedures as desktop instruments. Indeed, few archaeologists currently using PXRF instrumentation even analyse a standard during each group of runs. It is impossible, then, to determine whether the results are accurate (see

Glascock's discussion in Chap. 8). Nazaroff and Shackley (2009) recently combined a sample size study for PXRF, similar to the one for EDXRF discussed in Chap. 3, as well as a comparison of the results of the analysis of the same source rocks for the Antelope Creek locality at Mule Creek, New Mexico between a Bruker PXRF, the Berkeley empirically calibrated NITON PXRF and the Berkeley Quant'X (Shackley, 2005). The results were presented at the 2009 Geological Society of America meetings in Portland, Oregon: we can see that there are errors in the Bruker by a factor of 2–3, although some of the incompatible elements are very close. The number of elements from the Bruker analysis that are close to the source standard data is not necessary and sufficient to assign to source, or certainly separate the various localities like the multiple events at Mule Creek (see Shackley, 2005; Nazaroff and Shackley, 2009). The Bruker PXRF was not empirically calibrated, although Bruker claims it was "calibrated", but it appears that this was a fundamental parameter calibration.

The current vogue in PXRF in American archaeology seems to be (probably based on the instruction from Bruker and others), that all that is needed is to compare the count determined spectra between two or more obsidian artifacts and the analyst can discriminate whether they are from the same source after normalization – a decidedly qualitative and observer-based technique. There are at least two very crucial problems with this technique. First, it assumes that the analyst's judgement is correct, leaving one open to significant observer error. In this context, one must have at least a complete source database for comparison, just as in quantitative analyses. Second, and much more important, even with an adequate database of sources, I have seen at least three pairs of sources, quite distant from each other, that overlap exactly even after normalization, one misassignment by an industry representative touting this technique! The source pairs are: (1) Pachuca and La Joya in Mexico (see Glascock here); (2) Malad, Idaho and Cow Canyon, Arizona; and (3) Antelope Wells, New Mexico and Chihuahua and the newly located Los Sitios del Agua source in northern Sonora. These sources are hundreds, if not thousands, of kilometres apart, and I doubt that any archaeologist could confuse the two. Malad and Cow Canyon require a rather precise barium measurement to discriminate an element currently measured poorly with most portable instruments, which do not acquire much over 45 kV and thus make it difficult to determine that element.

In 2007, during Berkeley's Archaeological Petrology Field School in New Mexico, we analysed two very small ($\leq$10 mm) pieces of obsidian debitage from the Mockingbird Gap Clovis Site near Socorro, New Mexico at or near the time of excavation (Huckell et al., 2007). Table 2.1 exhibits the analysis of these samples by the Berkeley empirically calibrated Niton in the field, and the two samples on the Berkeley ThermoSpectrace QuanX, the same instrument used in the Craig et al. (2007) study. Both studies, because the Mount Taylor (Grants Ridge) source is an unusual mixed-magma source with high Y (not available with the Niton as purchased) and Nb, could assign these small samples to that source in the field (see Shackley 1998a, 2005). Other regional sources such as Vulture or Superior with relatively low mid-Z numbers may not be as easily discriminated (Shackley, 1995,

Table 2.1 Elemental concentrations for the analysis of two approximately 10-mm diameter obsidian debitage from the Mockingbird Gap Clovis Site, southern New Mexico with the Spectrace QuanX EDXRF and the Thermo Niton PXRF

Sample/ instrument	Mn	Fe	Zn	Rb	Sr	Y	Zr	Nb
419 (QuanX)	920	8311	156	519	7	69	103	181
419 (NITON-PXRF)	987	7700	171	486	13	nr	106	158
293 (QuanX)	856	7798	180	490	5	79	101	172
293 (NITON-PXRF)	656	6029	170	436	7	nr	100	nr
Mean Grants Ridge ($n = 15$); $\pm$ = SD	849 $\pm$ 64	8302 $\pm$ 385	154 $\pm$ 12	570 $\pm$ 29	4 $\pm$ 1	76 $\pm$ 3	119 $\pm$ 4	198 $\pm$ 6

PXRFAll Niton measurements made with the Niton sample holder. Mean "Grants Ridge" data are from the analysis of source standards with the Spectrace QuanX at Berkeley (see Shackley, 1998a)

2005). Indeed, a closer inspection of the data suggests that the empirically calibrated Niton is not as sensitive in the analysis of some elements particularly Mn, but again still exhibited an analysis that would allow for assignment to source. This is remarkable given that desktop EDXRF instruments are only useful down to about 10-mm samples with volcanic rocks as discussed earlier (see Davis et al., 1998, and Chap. 3, here; and Lundblad et al., 2008). Sample size is certainly an issue in PXRF analysis of volcanic rocks as in desktop instruments, but there is yet to be a systematic study similar to the Davis et al. (1998), Lundblad et al. (2008), and Chap. 3 studies with PXRF instruments, except for the Nazaroff and Shackley (2009) study that has been criticized by the PXRF industry.

This volume is not dedicated to PXRF; the Potts and West (2008) edited volume does that quite well. Liritzis and Zacharias here do a remarkable job as a 2009 example offering a review of the recent literature, and derive similar conclusions to the Craig et al. (2007) study and Williams-Thorpe's chapter in the Potts and West (2008) volume (see also Phillips and Speakman, 2009; and Pessanha et al., 2009). Like any emerging technology, PXRF will rapidly become more refined, and someday may replace most desktop systems. However, with the low-energy input, it may be a while before such a small "gun" captures all the elements between Na and U with the same instrumental precision as high-input desktop instruments (Shackley, 2010).

The Physics and Instrumental Technology of XRF

As noted in the first chapter, we endeavour to make the use of XRF in archaeology and geoarchaeology as transparent and understandable as possible for archaeologists. There are a number of texts and papers devoted to a more specialist discussion of XRF, mainly for physicists, chemists, and engineers who are more interested in the theory and, less so, in the method of XRF, but not as one of the many tools used by archaeologists to explain the past (Bertin, 1978; Franzini et al., 1976; Giaque et al., 1993; Jenkins et al., 1981, 1999). More recently, many web sites have discussions of XRF both commercial and academic (see especially http://www. learnxrf.com/). The remainder of this chapter is devoted to a rather in-depth discussion of XRF for archaeologists, many of who do not, unfortunately, have an extensive background in the physical sciences, especially those in the United States (Goldberg, 2008; Killick, 2008). Taken with the chapters here and the glossary in the back of the volume, we hope that a better understanding of the "black box" that is XRF will be easily grasped. Most of the chapters also discuss XRF for their particular application, from basalt or obsidian analyses with EDXRF, WXRF and NAA to the vagaries of PXRF. Again, for those who would like more depth, I refer you to Jenkins (1999) second edition of *X-Ray Fluorescence Spectrometry*. Most of those terms that are in *italics* are also defined in the glossary.

Theory and Derivation of XRF

X-rays are a short wavelength (high energy-high frequency) form of electromagnetic radiation inhabiting the region between gamma rays and ultraviolet radiation. The XRF method depends on fundamental principles that are common to several other instrumental methods involving interactions between electron beams and X-rays with samples, including, X-ray spectroscopy (e.g. SEM – EDS), X-ray diffraction (XRD) and wavelength dispersive spectroscopy (microprobe WDS).

The analysis of major and trace elements in geological materials by XRF is made possible by the behaviour of atoms when they interact with radiation. When materials are excited with high-energy, short wavelength radiation (e.g. X-rays), they can become ionized. If the energy of the radiation is sufficient to dislodge a tightly-held inner shell electron, the atom becomes unstable and an outer shell electron replaces the missing inner electron. When this happens, energy is released because the inner shell electron is more strongly bound compared with an outer one (Fig. 2.1). The emitted radiation is of lower energy than the primary incident X-rays and is termed fluorescent radiation, often called *fluorescence* in the vernacular. Energy differences between electron shells are known and fixed, so the emitted radiation always has characteristic energy, and the resulting fluorescent X-rays can be used to detect the abundances of elements that are present in the sample.

The Spectrum and Spectral Lines – The Electron Configuration of the Elements

This process of displacement of an electron from its normal or ground state is called *excitation*. The atom can return to the non-excited state by various processes, one of

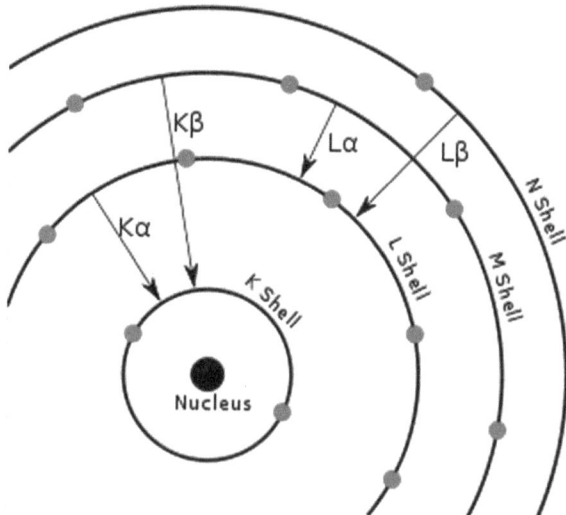

Fig. 2.1 Schematic view of orbital transitions due to X-ray fluorescence (see also Fig. 8.1 for an alternate view)

which is most important in XRF (Jenkins, 1999:55). When a substance is irradiated with high energy X-rays, electrons ejected from the atom produce an ion. The orbits or shells of an atom are called and read by the software as K through O lines (see Fig. 2.1). The K line transition is where the K electron moves out of the atom entirely and is replaced by an L line electron. Only the K and L lines are technically measurable with non-destructive XRF. X-rays of highest intensity from these transitions are called alpha transitions and are the transitions measured in all the XRF analyses in this volume (see Figs. 2.1–2.3). An L shell e-transition fills a vacancy in K shell and emits Kα1/Kα2 radiation. This is the most frequent transition, hence yielding the most intense and easily measured peak, the Kα peak. These Kα transitions are formed in doublets Kα1 and Kα2. Lα lines sometimes measured for those high Z elements such as Ba, particularly in WXRF instruments, are the result of M orbit transitions to L orbits. In Chap. 3, Davis et al. discuss the use of the Lα line in the analysis of Ba with the [241]Am gamma ray source in the retired Spectrace 440 EDXRF instrument at Berkeley. However, as also discussed in that chapter, most EDXRF labs today use very high energy ($\approx$50 kV) to excite the Ba atoms in volcanic rocks and measure the Kα peaks. While this does not yield the accuracy of using a gamma ray source, it avoids the issues of gamma radiation problems in storage and use, and the Ba elemental concentrations are generally accurate enough for most source assignments. Parenthetically, PXRF instruments are measuring Ba with much lower electron voltage. It remains to be seen what level of instrumental precision is offered in these instruments for the high atomic (Z) numbers.

Elemental Interference During XRF Analysis

While the measurement of fluorescent peaks in XRF seems straightforward, there are a number of interference issues that must be accounted for in all analyses. Many natural rocks consist of several different minerals of highly variable composition and structure. Even natural glasses that are amorphous with no crystalline structure are mixtures of a wide variety of chemical elements. This variable composition causes rocks to affect the behaviour of photons in highly complex ways. These effects on light translate directly to complexities in interpreting the fluorescence radiation that is detected in the XRF spectrometer. The complexities are collectively known as *matrix effects* which can be subdivided into overlap effects and mass absorption effects. The matrix effects on element i are the combination of mass absorption effects and overlap effects exerted on element i, by all coexisting elements j.

In Figs. 2.2 and 2.3, the spectra for the obsidian USGS standard RGM-1 prepared as a pressed powder pellet, the overlap between elements are readily observable. As an example, in Fig. 2.2, the mid-Zb analysis and ratio to the Compton scatter, Rb Kβ1 overlaps the primary Nb Kα1 peak. These are called peak overlap or interference effects, and in early geoarchaeological studies using WXRF in particular, the software was not available to "strip" overlapping peaks from others by deconvolution

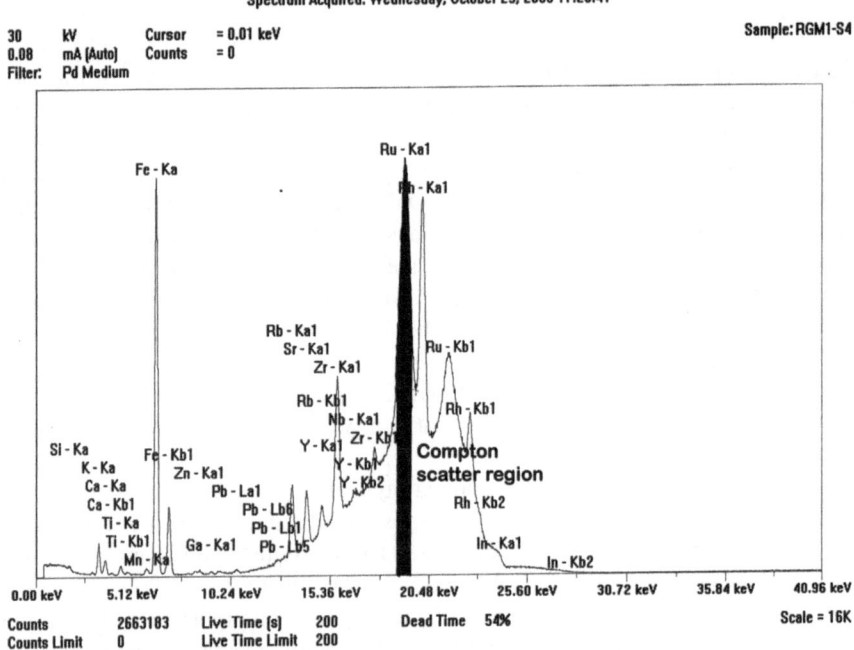

Fig. 2.2 Spectrum of the ThermoScientific Quant'X mid-Zb analysis of USGS RGM-1, showing the Compton scattered "hump" and that portion of the region under the Ru peak (*blackened*) used for peak ratioing

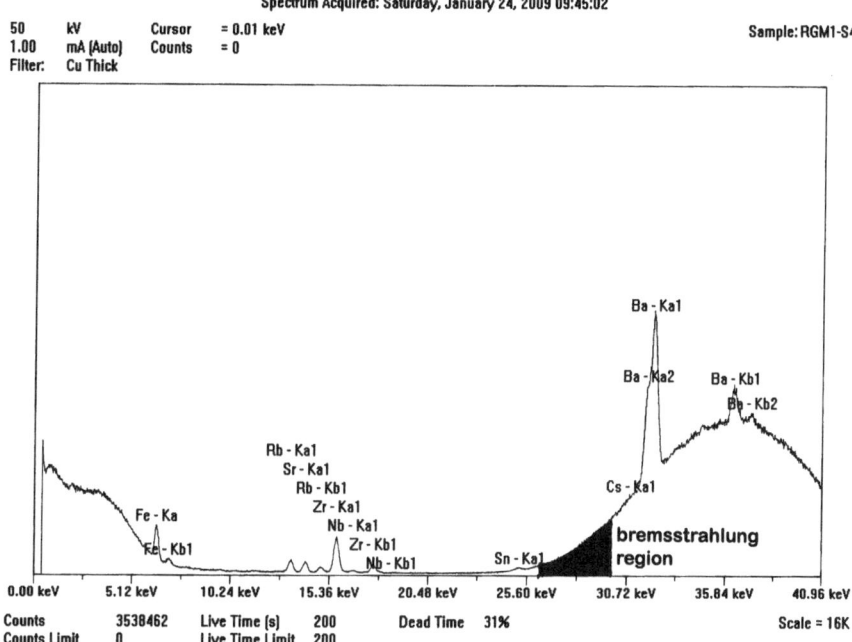

Fig. 2.3 Spectrum of the ThermoScientific Quant'X high-Za analysis of USGS RGM-1, showing the bremsstrahlung region and that portion of the region (*blackened*) used for peak ratioing

(Hughes, 1983, 1984; Ivanenko et al., 2003; Shackley, 1988, 1990). In this case, the peak heights were ratioed to others providing a semi-quantitative analysis. These semi-quantitative analyses were found to be problematic, because it is possible for two or more volcanic sources to have the same ratio of the selected elements, but different concentrations. In the North American Southwest, this is the case with the Antelope Wells (El Berrendo) obsidian source in southwestern New Mexico/ northwestern Chihuahua and the recently discovered Los Sitios del Agua source in north-central Sonora, sources hundreds of kilometres apart (Martynec et al., 2010; Shackley, 1988, 1995, 2005). They are both *peralkaline* or mildly peralkaline glasses with very different elemental concentrations, but when plotted on the older ternary systems, they plot at the same point. Quantitative analyses (weight % and parts per million measurements, PPM) effectively solve this problem.

Not apparent in the spectra are *mass absorption effects*. Mass absorption effects result from fluorescence radiation being absorbed by coexisting elements (causing reduced intensity), or enhancement of fluorescence radiation due to secondary radiation from itself or coexisting elements (causing increased intensity). In many cases the effects can be effectively eliminated by proper sample preparation in pressed powder or fused disk samples, but corrections can be made in any case even when analysing samples non-destructively.

Table 2.2 X-ray fluorescence concentrations for selected trace elements of RGM-1 pressed powder pellet ($n = 35$ runs), and whole rock flake from original USGS Glass Mountain boulder ($n = 12$)

SAMPLE	Ti	Mn	Fe	Rb	Sr	Y	Zr	Nb	Ba	Pb	Th
RGM-1 (Govindaraju, 1994 recommended)	1600	279	12998	149	108	25	219	8.9	807	24	15.1
RGM-1 (USGS recommended)[a]	1619 ± 120	279 ± 50	13010 ± 210	150 ± 8	110 ± 10	25[b]	220 ± 20	8.9 ± 0.6	810 ± 46	24 ± 3	15 ± 1.3
RGM-1, pressed powder (this study, $n = 35$)	1563 ± 60	302 ± 14	13116 ± 308	151 ± 3	106 ± 3	25 ± 2	219 ± 5	9 ± 2	869 ± 61	26 ± 2	16 ± 3
RGM-1, flake from original USGS boulder ($n = 12$)	1568 ± 44	311 ± 11	13306 ± 33	153 ± 2	113 ± 2	25 ± 1.5	230 ± 4	9 ± 2	842 ± 14	23 ± 1.5	15 ± 3.4

Plus or minus values represent first standard deviation computations for the group of measurements. All values are in parts per million (ppm) as reported in Govindaraju (1994), USGS, and this study. RGM-1 is a U.S. Geological Survey obsidian standard obtained from Glass Mountain, Medicine Lake Highlands Volcanic Field, northern California. All samples analysed on the ThermoScientific Quant'X EDXRF spectrometer at Berkeley

[a]Ti, Mn, Fe calculated to ppm from weight % from USGS data

[b]Information value

Today, matrix effects particularly mass absorption and overlap effects are eliminated by stripping routines that calculate the intensity of each element of interest and "strip" them from overlapping elements. This is a tremendous advancement in the software routines in XRF and invisible to the analyst (see Ivanenko et al., 2003; Lachance and Claisse, 1994; McCarthy and Schamber, 1981; Schamber, 1977).

Issues of Practical Matrix Effects

Practical matrix effects here are the practical issues of analysing two different matrices, such as pressed powder standards and analysed whole rock. We have worried about this for a number of years. Most of us using EDXRF instruments use pressed powder pellets of international standards for calibration and checking various analytical runs. The question, posed by some, is whether setting up calibration routines with pressed powder pellets for the analysis of whole rock samples is skewing our results (Mike Glascock, personal communication 2009). One of the more common standards used is RGM-1, a U.S. Geological Survey obsidian standard from the Glass Mountain obsidian flow of the Medicine Lake Highlands of northern California. A 200 kg single block of obsidian was collected by USGS and powdered. It is sold, or was sold (there is a newer RGM-2) to anyone desiring the standard. I was able to procure a whole rock flake of RGM-1 from the original boulder from USGS, and using our calibration routine based on pressed powder pellet standards (see Appendix), derived very similar results from the flakes as the powdered standards (Table 2.2). We had also tested this years ago as reported here in Chap. 3. I think this is a non-issue, at least for obsidian.

Evaluating Spectra: Compton, Bremsstrahlung and Other Spectral Issues

Evaluating merely the elemental spectra is only part of the work performed by the XRF analyst and the software available. The influence of the background radiated, often called "scatter" in the XRF vernacular, is important to understand, and is useful in determining the quantitative composition of a sample through "ratioing".

Backscatter

Some of the X-rays strike the sample itself (i.e. an obsidian artefact) and are scattered or reflected directly into the detector. For instrumental XRF, this is one form of scatter that is stripped from the analysis even though the quantity of backscatter is mass dependent (Jenkins, 1999, 24–25).

Rayleigh Scatter

The Rayleigh scatter, also called elastic or coherent scatter, occurs as a result of a portion of the X-rays from the tube bouncing off the atoms without producing fluorescence, but it occurs as a source peak in the analysis. In essence, the high-energy X-rays directed at the atoms in the sample are partially redirected into the detector from the atoms directly. In Figs. 2.2 and 2.3 they are part of the elemental peaks seen in the spectrum but are not measured in the analysis, just as with backscattering energy.

Escape Peaks

In a gamma or X-ray spectrum, the peak due to the photoelectric effect in the detector *escapes* from the sensitive part of the detector. In XRF systems with Si(Li) detectors (as in most systems discussed in this volume), as X-rays strike the sample and promote elemental fluorescence, some Si fluorescence at the surface of the detector escapes, but it is not collected by the detector. The result is a peak that

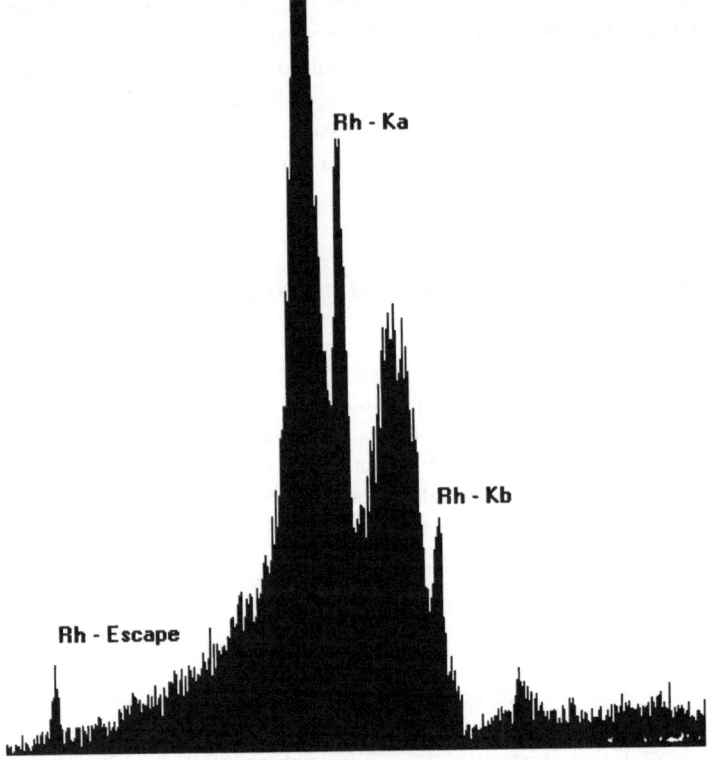

Fig. 2.4 A Compton scatter showing the Rh escape peak at 1.74 keV

appears in the spectrum, at: element keV – Si keV (1.74 keV; Fig. 2.4). In most archaeological applications, these escape peaks have virtually no relevance, but for quantitative geological applications, they are worthy of notice and calculation (Buras et al., 1974). Again, modern software virtually eliminates these peaks as an issue.

The Compton Scatter

An important portion of the spectra in the generation of quantitative elemental analyses in non-destructive XRF is the Compton scattered peak (Figs. 2.2 and 2.3). Also called incoherent or inelastic scattering that occurs when an X-ray photon collides with a loosely bound outer electron. The electron recoils under the impact, removing a small portion of the energy of the primary photon, which is then deflected with the corresponding loss of energy, or in WXRF increase of wavelength. There is a relatively simple relationship between the incident λo and incoherently (Compton) scattered wavelength λc:

$$\lambda c - \lambda o = 0.0242(1 - cos\psi). \tag{2.1}$$

ψ is the angle over which the X-ray beam is scattered which in our spectrometers is equal to 90°. Since the cosine of 90° is 0, there is generally a fixed wavelength difference between the coherently (elemental in our case) and incoherently scattered lines equal to about 0.024 Å. As noted by Jenkins: "This constant difference gives a very practical means of predicting the angular position of an incoherently scattered line. Also the incoherently scattered line is much broader than a coherently scattered (or diffracted) line because the scattering angle is not a single value, but a range of values due to the divergence of the primary beam" (1999:12). This is crucial in both non-destructive EDXRF and WXRF applications where the artifacts have varying sizes and surface configurations. In tube driven XRF instruments, the Compton scattered peak or background is produced by the target (Rh targets in this volume) and produces those large Rh peaks, a portion of which is used for ratioing to the elemental peaks (Fig. 2.2). So, by ratioing to this Compton scatter, which is in essence a direct reflection of the mass of an object, a 10-mm piece of obsidian biface thinning debitage produced from the Sierra de Pachuca source in Hidalgo, Mexico, will have, through a non-destructive EDXRF analysis, the same composition as the much larger biface from which the flake was struck. Without the knowledge and application of the Compton scatter, we could not analyse archaeological material non-destructively.

The Bremsstrahlung Region

Bremsstrahlung (or continuum or continuous) radiation is the German for "breaking radiation", noise that appears in the spectra due to deceleration of electrons as they strike the anode of the X-ray tube. This is also frequently called the *background*. For the heavy incompatible elements such as Ba which are important for discriminating volcanic rock sources, bremsstrahlung scattering does appear at the heavy end of the spectrum. The bremsstrahlung radiation is produced by the tungsten (W) anode in the X-ray tube and (conveniently) provides a region for ratioing for the heavier elements as shown in Fig. 2.3. In the EDXRF case or in WXRF where ratioing is used for non-destructive analyses, a region must be chosen that does not contain any element of interest (see Davis, 1994; Chap. 3 here). So a region on the bremsstrahlung scatter is chosen for comparison and more specifically the region between Sn Kα1 and Cs Kα1 is used at Berkeley for ratioing for Ba analysis (see Fig. 2.3).

The Influence of "Background"

The term "background" used in the XRF vernacular is the bremsstrahlung radiation as discussed above. Since scatter increases with a decrease in the atomic number of the "scatterer", backgrounds are much higher for low average atomic number specimens (or an analysts attempt to analyse elements of lower atomic number) than the background for higher atomic numbers. As an approximation, the background in XRF varies as $1/Z^2$. In essence, this is why Rh target EDXRF instruments and XRF in general are unable to reliably analyse below about $Z = 11$ (Na). The background in the low atomic number elements is relatively too high for XRF to deal with the counting errors vs. background. With an EDXRF Ag target instrument it is possible to get down to $Z = 9$ (F), but not easily. The term "mid-Z" elements in XRF vernacular relate to those elements from about $Z = 19$ to $Z = 41$, and because the background effect is relatively inconsequential, those elements are analysed readily with predictable error.

In sum, the various spectra in the generation of X-rays have various uses in archaeological XRF analysis. X-ray tubes with tungsten anodes and rhodium or silver targets produce continuous radiation (W anode) and incoherent scattering (Ag or Rh targets) that have great utility in concert with modern software and precision detectors in providing good instrumental precision and analytical accuracy solving some of our more interesting and important problems in archaeology today.

XRF Hardware and Software

The foregoing discussion of XRF theory and method is only part of the story. Those of us involved in the XRF analysis of archaeological materials make hardware decisions for any number of reasons: (1) budgetary limitations obviously; (2) the history of certain hardware at our institutions or those where we learned XRF; (3) the types of materials analysed; (4) and institutional agreements that require purchasing from only certain suppliers.

In this section, the differences between EDXRF and WXRF, the important components of these instruments, and useful issues of instrument acquisition, calibration and the derivation of the results that we send to archaeologists are discussed.

Acquisition Condition Selection

Choosing optimal acquisition conditions for XRF analysis is a complex and critical part of the work, and to a certain extent part of the "art" of XRF spectrometry. Selecting the proper acquisition conditions can mean the difference between measuring an element at PPM levels, or not seeing it at all. There are two fundamental principles that must be met to achieve optimal analysis conditions.

There must be a significant source peak above *absorption edge* energy (the upper limit of the K or L radiation) of the element of interest. This may be either the K or L line depending on which one is appropriate as discussed above. The closer the source energy is to the absorption edge, the higher the intensity and sensitivity (counts per s/ppm) will be for the element of interest.

The other fundamental principle is that the background X-rays within the region of the elements of interest should be reduced as much as is practical. The difficulty is that these two principles work in opposition to each other, as the best sensitivity is often achieved when the background is highest, and the background is lowest when the sensitivity is poor – the weakness of XRF analysis. Added to this is the fact that the best theoretical detection limits are achieved when the sensitivity is highest, while the net count rate extraction, matrix corrections and long-term analytical stability are best when the background is lowest. Optimal analytical performance is achieved by finding the best compromise between these two principles, given the instrument hardware. In modern instruments, these issues are partly corrected by very real increases in the sensitivity of modern detectors and a shift from analogue to digital connections between the instrument and the computer and software.

Elements of Interest

The first step to setting up a XRF analysis is determining the elements of interest. If a sample or rock type has never been analysed for every conceivable element, the odds are high that it contains something that we might not expect. Some samples come into a laboratory as complete unknowns, such as obsidian artifacts from a region unfamiliar to analysts in the lab. For example, many of the rhyolite centres that produced obsidian in the Rift Valley in East Africa contain relatively high concentrations of Zn, much higher than obsidian in the rest of the world (Negash and Shackley, 2006; Negash et al., 2006). Zinc becomes one of the best discriminating elements in the region, particularly those sources in Ethiopia and to a certain extent the Near East, but has little utility in other regions.

If a sample is not well characterized, it is a good idea to perform a qualitative examination of the material using three or more acquisition conditions, designed to cover high, medium, and low energy ranges. Qualitative acquisition conditions will be covered below. Alternatively, a multivariate statistical analysis such as principal components analysis can isolate those elements of interest that are best discriminators in the region (see Glascock et al., 1998).

Source Selection – Isotopes

Isotopes are the simplest source to configure. Select a source that emits X-rays that are closest to and immediately above the absorption edge energy for the element of interest. To avoid problems with high background, the element of interest peak should be at least 2–3 times the *FWHM* (full width – half maximum of the peak) detector resolution away from the source peak. Davis et al. in Chap. 3 discusses the use of the [241]Am source at Berkeley in the 1990s for acquisition of barium that can be problematic with X-ray tube analyses as discussed above. FWHM is an expression of the extent of a function, given by the difference between the two extreme values of the independent variable (this would be the elements of interest) at which the dependent variable is equal to half of its maximum value (see X-ray filters below). In EDXRF this is also called resolution, calculated as the distance in electron volt between left and right sides of the peak at half of its maximum height, or more simply the peak width at half its height.

Source Selection – X-ray Tubes

X-ray tube selection is often done by the manufacturer without much input from the customer, but there are some selection rules that are useful. X-ray tubes emit a broad bremsstrahlung spectrum from 0 to X KeV, where X is the accelerating voltage of electrons that strike the metal target in the X-ray tube. The peak intensity in the bremsstrahlung spectrum is at roughly half the maximum energy. The X-ray tube also emits line energies that are characteristic of the target element, so target selection is usually based on selecting a target that will provide optimal excitation for the most important elements of interest. Alternatively, the selection is based on having a line energy that does not increase the background in the region of any important element. Since the amount of X-ray flux is proportional to the atomic number of the target element, anodes such as W are also selected on the basis of having the highest total flux. Targets can be any of several high-melting point metals. Common target choices include Sc, Ti, Cr, Fe, Co, Ni, Cu, Y, Zr, Mo, Rh, Pd, Ag, W, and Pt. Rhodium is the most common among the ones used in the EDXRF discussions in this volume, but as discussed above, Ag targets can aid in the acquisition of the lighter elements and can be constructed as easily as tubes with

Fig. 2.5 Schematic of a portion of a typical end-window X-ray tube

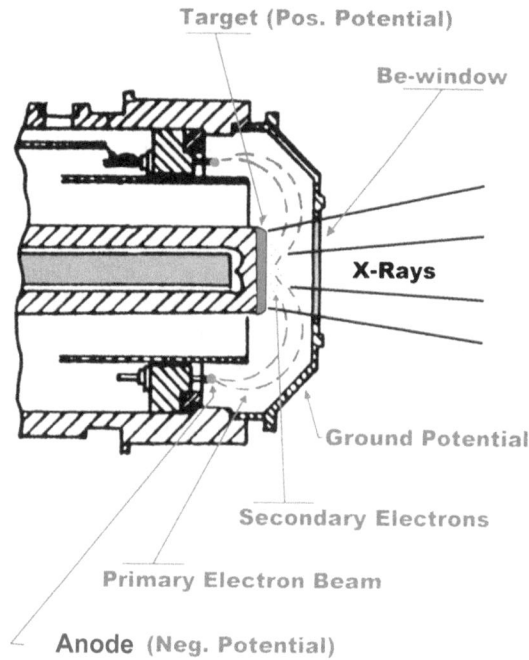

Rh targets (Richard Hughes, personal communication, February 2009; Franzini et al., 1976).

Multiple target tubes are often used in WXRF instruments, and they allow the operator to select the target on a per element basis, using the rule that the target with line energy immediately above the absorption edge is the one selected, providing it is at least 2–3 times the FWHM detector resolution away from the element line. If there is no target available with an emission line above the element of interest, the analyst or the software selects the highest atomic number target available to maximize the total X-ray flux from the tube.

Most modern EDXRF tubes are composed of a tungsten anode and an Rh or Ag target, and they are all end window tubes, such as the ThermoScientific products. Modern WXRF systems employ a variety of different tube configurations, and as mentioned below, multiple target and collimation configurations (Fig. 2.5)

X-ray Tube Filters

Filters are frequently placed in the X-ray path between X-ray tube and sample in order to modify the shape of the source spectrum. Filters can be made of any element that can be formed into a stable solid or film. They are usually metal or plastic although plastic filters deteriorate under prolonged bombardment by X-rays.

The key to the function of filters is the filter element's absorption edge energy. The filter readily absorbs source X-rays immediately above the absorption edge while those below the absorption edge are transmitted. Very high-energy X-rays are also transmitted. This produces a low background valley immediately above the filter's absorption edge that is crucial for analysing elements in the energy range beginning 2 FWHM detector resolution widths above the absorption edge. In the Thermo-Scientific EDXRF spectrometers, seven filters are used for analysis of elements between $Z = 11–92$. As an example, for $Z = 37–42$ (Rb–Mo) a 0.06-mm Pd filter is optimal (called by Thermo the "Mid Zc" region). Palladium ($Z = 46$) is 3Z past Mo in the periodic table and over 2 FWHM past Mo. However, with proper calibration it is possible to get relatively accurate numbers for Ti–Nb, plus Pb and Th with the medium Pd filter with obsidian – a very homogeneous substance (see Shackley, 2005: Appendix). For the analysis of Ba, a 0.559 thick Cu filter is used. This is well beyond the FWHM, because the tube is operated at such a high energy, typically 50 kV and 0.5–2 mA.

Filters also fluoresce their own characteristic line energies, which combine with the bremsstrahlung that is transmitted by the filter to a region that resembles a right triangle (see Fig. 2.3). This secondary filter fluorescence peak can be used as a source peak for elements that are about 3 atomic numbers or more less than the filter element, as discussed above.

Another type of filter is a neutral absorbing filter such as aluminium or cellulose. These filters are intended to filter lower energy source X-rays in order to reduce the background in the region of the element of interest. A thin neutral density filter may be useful for measuring elements like S, or P with a Rh, Pd, or Ag target X-ray tube, while thicker Al filters can eliminate these target peaks entirely creating a source that is good for analysing X-rays (Kα lines) between 2 and 10 keV (P–Ge elements). In the Thermo instruments, no filter is used for the very lowest end of the spectrum ($Z = 11–16$; Na–S). In this case, the energy is so low that any filter would reduce the very low fluorescence to near zero even in vacuum. In advanced PXRF instruments, a number of filters will be combined (i.e. aluminium, palladium) to enable a broad range of elemental acquisition. To my knowledge, this has not been tested experimentally, or at least reported.

Voltage Settings (kV)

Once the type of acquisition is determined, choosing the optimal X-ray tube voltage is the next step. Because of the broad energy distribution created by the X-ray generation process of an X-ray tube, the optimal high voltage is usually 1.5–2 times the absorption edge energy of the highest energy element in the acquisition. This element may be an element of interest but is more commonly the X-ray tube target, secondary target (not discussed here) or filter material. It may also be the K absorption edge Kα, Kβ, or the L absorption edge depending on which lines are being excited for analysis. If optimal deadtime (periods at which counts are not taken because the detector is busy with earlier detected X-rays) or count rates cannot be achieved at the 1.5× value due to current limitations of the tube or high voltage power (usually near 50%), then the high voltage should be increased until they are close to that level. In modern systems discussed here, the power setting (kV and mA) is generally set by the software, particularly for milliamperage (mA), the "push" or current of the voltage.

X-Ray Tube Target Excitation

If a characteristic target line is used to excite some elements, then the analyst selects a voltage 1.5–2× its absorption edge. For example, in the Thermo instruments, if Rh K-lines are used as the excitation source and its Kαβ energy is 23.224 keV, then voltage is set in the 35–45 keV range. In the Thermo instruments, this is usually at the lower end (i.e. about 35 keV) for the mid-Zc region, in part because the digital pulse processors are much more efficient now.

When the bremsstrahlung continuum acts as the exciting radiation, then the high voltage should be 1.5–2× the highest energy element of interest excited in that analysis condition. For example if we are measuring the Ba Kα line with its 37.410 keV absorption edge, then 56 kV potential is recommended, but since the Rh target maximum is 50 kV, that is used. This is one reason that tube acquired Ba measurements are subject to such varying intensities, as found in the comparison with the use of the Am source (see Chap. 3). By using the thick Cu filter, some of this problem can be eliminated such that the error rates in the analysis of Ba are diminished.

Current (mA)

There is one simple rule for setting the current; measure the count rates or deadtime at the lowest current setting, usually 1 or 10 mA. The X-ray flux from the tube increases in direct proportion to the current so it is simple to extrapolate the needed current. The detector response is not quite linear, so an estimate of the current is needed to reach the instrument manufacturers specified optimal counts, and adjust the current upward. The reason this is done is that detectors do not respond well to excessive count rates, and if the maximum count rate has been unknowingly exceeded, the instrument will fail to make a proper measurement: essentially, when deadtimes get too high, peak resolution degrades. This is partly detected by the deadtime and in the Thermo WinTrace™ software, it is noted in red on the monitor when it reaches a very high or very low point. Optimally, deadtime should be around 50%. By changing the current, the deadtime can be optimized.

In general, a higher current is required when operating at low voltages (less than 10 keV). In some cases, an instrument may have too high a count rate at the minimum current. In such case, the high voltage must be reduced or a different filter or collimator selected. Most modern EDXRF instruments can set the optimal current for a given voltage that is sufficient for most applications. It is still a good idea to experiment with varying tube conditions, particularly if the range of elements of interest is outside that recommended by the manufacturer.

Tube Collimators

Collimators are another option in XRF instruments. In EDXRF systems, they usually have a single hole in the middle and can vary in size from 25 μm to several millimetres. Collimators are usually selected when small spot sizes are needed either because the sample is small or a specific point of interest on a sample is small. As mentioned above collimators are also used to reduce the X-ray intensity in some cases when sample size may be large and deadtime is too high, such as large obsidian or basalt bifaces or other large tools (see Lundblad et al., 2008; and Chap. 4). The collimator used on the Berkeley Quant'X is 8.8 mm in diameter and creates a 28 mm circular irradiated area on the samples. For most analyses, this is fine, but when the sample is very small, it is necessary to use a smaller collimator, 3.5 mm in this case. I have not found it to require a different calibration. It is important to remember that while *infinite thickness* (discussed in Chap. 3) can be a problem in XRF when X-ray penetration depth exceeds the thickness of the artefact, it is essentially the mass in general that is important in whole rock analyses.

Atmosphere

Air readily absorbs low energy X-rays, particularly for elements below titanium in the periodic table. Since argon makes up 1% of the composition of air and has an absorption edge below K and Ca, air also absorbs X-rays from those elements. It is common when analysing low Z elements to change the atmosphere by purging the chamber with helium (for liquids) or evacuating it entirely for solids. Lundblad et al. discuss in Chap. 4 and (2008) that all their analyses are conducted in a vacuum because of the dominance of the analysis of heterogeneous basalts non-destructively. It is necessary for low Z elements but is useful in their work for all elements of interest, even for those with $Z \approx 41$. Evacuating the analysis chamber is preferred when analysing all the low Z elements, since a vacuum does not absorb X-rays.

Purging is not recommended when analysing higher energy elements. Because the light element X-ray intensities are higher when the chamber is purged, there are fewer available counts for the heavy elements. When air is in the chamber, more current is required to achieve optimal count rates, and the net affect is improved sensitivity for the heavier elements. In this way the air functions as a neutral density detector filter that reduces light element intensities.

Count Time

The standard criteria for selecting count times are convenience and precision. Most analysts will use measurement times from 10 s to 10 min. Shorter count times 10–30 s are used more for qualitative scanning and sorting. The concentration range of the elements of interest is also important. Major elements in percent concentrations can be analysed in a minute or less, while minor elements at PPM concentrations may need to be analysed for 3–10 min or longer. The other and ultimately more important criterion is precision. Unfortunately measurement precision cannot be determined until after a calibration is complete, because only then is the calibration slope known. Most operators will use longer count times than necessary at first, in order to avoid having to repeat the calibration later.

Obsidian analysts in the U.S. who use EDXRF generally irradiate for 150–300 live seconds. However, as Giaque et al. found, increasing the counting time can increase precision, all other factors held equal (1993). One must evaluate the effort to achieve great precision over the accuracy needed to assign artifacts to source however (Shackley, 2002).

X-Ray Detection

Once a sample has been excited to fluorescence, a detector is used to convert X-rays into electronic signals which can be used to determine energy and intensity (number of X-rays) emitted from the sample. There are two types of detectors commonly used, the proportional counter used in WXRF and the semiconductor detector. The former is rarely used in archaeological applications.

The Si(Li) Detector

The Si(Li) semiconductor detector incorporates a silicon chip, which responds to X-rays by producing a charge at the detector output. This charge is converted into a voltage pulse which is then directed to pulse processing (Fig. 2.6). In the Si(Li) detector incident X-rays produce ionizations of the Si. The sensitive Si region is increased through the use of a process known as lithium drifting. Incident X-rays produce ionizations of Si in the sensitive regions of these detectors. The charge carriers are negative electrons and positive "holes", which are drawn to opposite ends of the detector due to the voltage bias applied across the silicon chip. Total charge collected within the semiconductor detector is directly proportional to the energy of the incipient X-ray, and is converted to a corresponding voltage amplitude through the use of the preamplifier and amplifier (see Fig. 2.6). For the

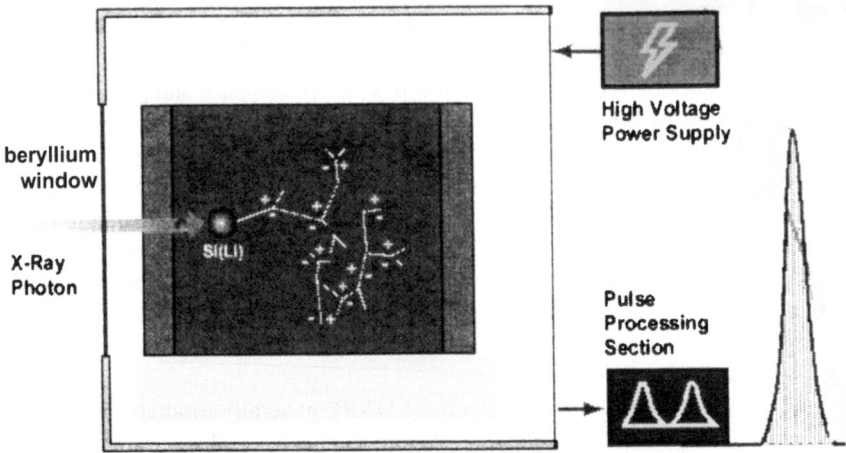

Fig. 2.6 Schematic representation of a Si(Li) detector in an EDXRF system

ionization to occur, the detector must be kept at very low temperatures and high vacuum, and it is sealed by a beryllium foil at the entry point.

Recently new germanium detectors and silicon drift detectors (SDD) without lithium are achieving better count rates and/or better resolution that Si(Li) detectors used most often in archaeological applications. These new X-ray detectors have yet to be applied in geoarchaeology, but will soon, perhaps solving some of the precision issues discussed above.

Pulse Processing

Charges produced in response to ionization in the detector are fed through a preamplifier and emerge from the detector output and need to be smoothed in the amplifier which gives them a pulse shape. Finally, these pulsed data are sent to the multi-channel analyzer (MCA), which converts the analogue pulses into channels. A channel is a memory location representing a small range of energies. As each pulse is digitized, it is stored in a channel corresponding to its amplitude (representing an X-ray energy level) and a counter for that channel is incremented by 1. The end result of these measurements is a collection of stored digital signals sorted by pulse height. These values are displayed graphically as a frequency distribution (histogram of energy vs. intensity) referred to as the *spectrum* (see Figs. 2.2 and 2.3), and further reduced through the calibration routine software into useable weight percent or parts per million data. At all stages in the pulse processing chain, proportionality between the detected X-ray energy, the analogue pulse amplitude, the digital signal value, and the corresponding channel number is strictly maintained.

Instrument Standards and Empirical Calibration

Students often ask, "How do you compare the results from one lab with another?" The answer is relatively simple. All laboratories, regardless of the type of instrument used, calibrate their instruments using the same international standards, mostly rock standards. These rock standards are supplied as powdered samples from mainly federal institutions from any number of countries, including the U.S. Geological Survey, the U.S. National Institute of Standards and Technology, and the Geological Surveys of Canada, France, Japan and South Africa (see Appendix). Additionally, during each analytical session, a selected standard is run with the unknown samples in order to determine whether the instrument is running within acceptable parameters, and just as important, so that other labs can evaluate the results compared to theirs.

For many years, the Berkeley lab has collaborated with those at the University of Missouri (Michael Glascock), the Geochemical Research Laboratory (Richard Hughes), and the Northwest Research Obsidian Studies Laboratory (Craig Skinner), and in some cases we may have to evaluate the data from one or another of the labs, and I have always found the data so similar to source data from my lab, that I can readily determine the source. The differences are within a few percent, often only 1%. Of course, both Hughes and Skinner use the same software and instrument, but even the NAA and XRF data of Mike Glascock at Missouri are within a percent or two for most elements, except for those elements that XRF does not measure well or vice-versa (see Chap. 8).

The instrument settings for the ThermoScientific Quant'X at Berkeley are in the Appendix, but just how does an XRF analyst set up a method to analyse volcanic rocks? The minor details between these different instruments and laboratories are slightly different, but the trajectory is essentially the same.

The EDXRF Quantitative Method Trajectory

As with any good research project, the research plan must be formulated first:

- What is the sample type – pressed powder, fused disk, whole rock?
- What are the elements of interest – major oxides, trace elements, rare earths?
- What level of precision is acceptable?
- What level of accuracy is necessary to assign to source?
- Conditions to be used- measurement time, voltage, amperage?
- What standards are available or necessary?

After the outline of research, the first step in EDXRF is to acquire the elements and generate elemental peak profiles. This is necessary for the peak-fitting algorithms that correct for overlap and background. Pure elements purchased from a number of suppliers in powder form are acquired by a software utility and incorporated into the method file. For lighter elements, and those that will measure L lines, the same voltage and filter settings should be the same.

The next step is to select the conditions and elements of interest. Most instrument software restricts the selection to only those that can be optimally acquired given the conditions selected. This eliminates many problems of interpretation.

The next step is to create a standards library. Standards libraries are separate files that contain standards used to calibrate the method. Separate standards libraries may be used for various methods or conditions. At Berkeley, we have separate standards libraries for the oxides using fused disk standards and for trace elements using pressed powder standards. This is a typical XRF strategy, but see Lundblad et al. (2008) for a non-destructive analytical strategy for basalt. An important consideration when selecting standards is that they should exhibit the entire range of variation expected from the rocks to be analysed. For obsidian (rhyolite), this would include (for strontium for example) a range between zero and around 500–600 ppm. It is rare for a rhyolite to be much above 200 ppm. In the Berkeley case, we also analyse basalt, dacite, and other rocks as well as ceramics, and so include standards from all the major volcanic rock groups, and the ceramic standard SARM-69 from South African Neolithic pottery (see Appendix).

After importing the standards for a given method, the instrument is calibrated using the expected (given) elemental concentrations (let us say ppm) from the standards vs. the calculated elemental concentrations (ppm). This is called *empirical calibration* as opposed to a fundamental parameters calibration. In this case a linear algorithm is used for calibration and all the standards are analysed as unknowns and the elemental data from each standard is plotted relative to a best fit regression line (Fig. 2.7). For the oxide fundamental parameters analysis, a quadratic regression may be used. Fundamental parameters calibrations use a variety of calculus functions to predict the expected values, although some are also based on a linear algorithm.

After the instrument completes the calibration routine, a table and regression scattergram is produced. The fit to the line can be improved by elimination of data from any standard until a best fit is achieved (Fig. 2.7). This could require a number of calibrations. Figure 2.7 exhibits the calculated vs. given concentrations for Sr in the Quant'X system. In this case, none of the standards have been removed. The 17 standards included in the analysis vary from near zero ppm Sr for JR-1 (40 ppm) and JR-2 (39 ppm), Japan Geological Survey rhyolite standards to nearly 1,300 ppm for BR-N, a Geological Survey of France basalt standard. The good fit of the standards on the regression line indicates that the analysed concentrations were at or near the given concentrations recommended by the various institutions or the Geostandards Newsletter (Govindaraju, 1994). The recommended data for each element of each standard is an arithmetic mean of data submitted by laboratories worldwide for those particular elements and standards.

The calculated values for various elements in the Geostandards Newsletter are used by most laboratories around the world, so each lab expects, for example, that RGM-1 USGS obsidian standard should yield an elemental concentration at about 108 ppm. Again, this shows that each instrument is properly calibrated and allows scientists from other institutions to evaluate interlab bias.

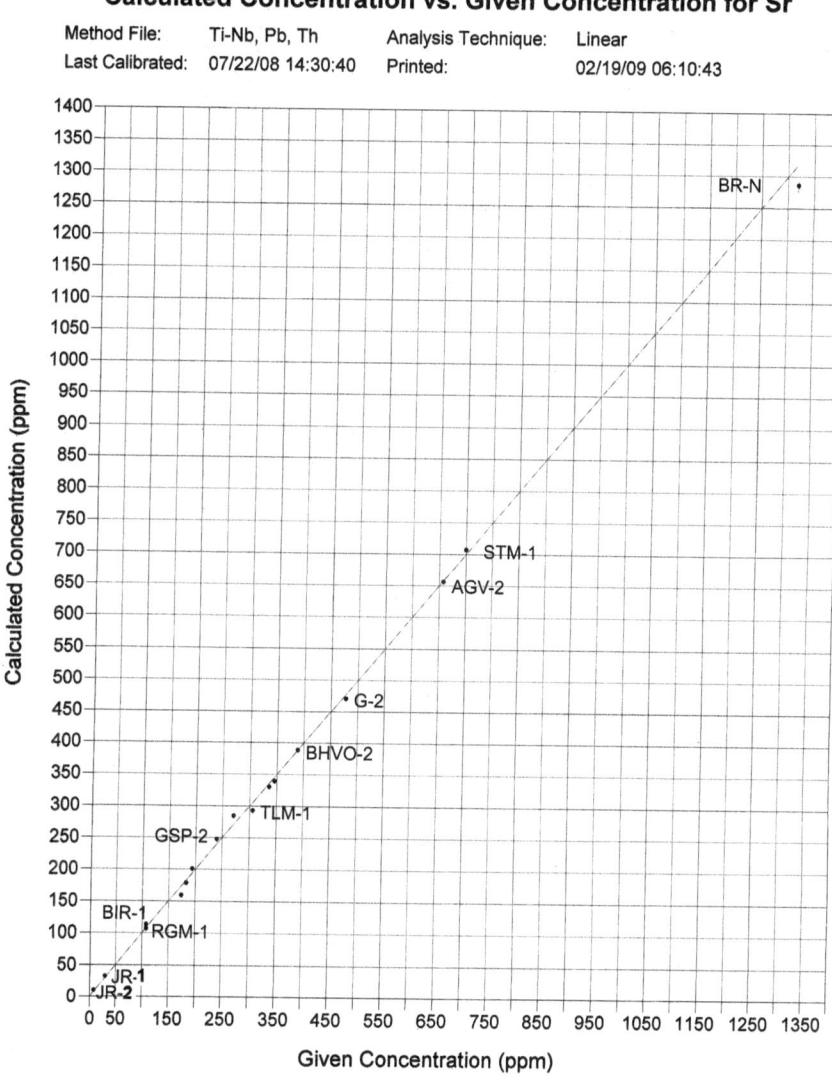

Fig. 2.7 Given (*x*-axis) vs. calculated (*y*-axis) calibration curve for Sr on the ThermoScientific Quant'X using 17 international standards (see Appendix for list of standards). Only selected standards noted. All are present

Qualitative Analysis

There is little archaeological literature covering XRF qualitative analyses, but it is a routine frequently used to determine the relative composition of a substance. Qualitative analysis is the detection of elements without applying rigorous quantification methods. We frequently use it to determine whether an element or set of elements are present. As mentioned above, a qualitative analysis can aid in determining which elements best discriminate a given obsidian source or sources within a geographic region. It is impossible to come up with a single acquisition condition that excites every element well with low background in every region of interest. To overcome this problem a qualitative analytical procedure will consist of three or more different sets of acquisition conditions. Three basic acquisition conditions cover the same low, medium, and high energy ranges discussed previously.

Figure 2.8 shows a qualitative analysis of the USGS RGM-1 standard. Compare with Figs. 2.2 and 2.3, quantitative analyses of the same standard, actually the same sample. This medium count qualitative analysis between about 1 and 20.48 keV does indicate the elemental composition of this obsidian. It does not tell you the quantity of each element but does show the elements in relative proportion given background. Note that the bremsstrahlung (continuous) background is readily observable under the labelled Rb–Zr elemental peaks (the hump-shaped region at ~10–19 keV).

Qualitative analyses like this are frequently used in environmental studies to determine the presence of elements of interest, such as lead in paints, mercury in water, and other contaminants. This is an application where PXRF is particularly

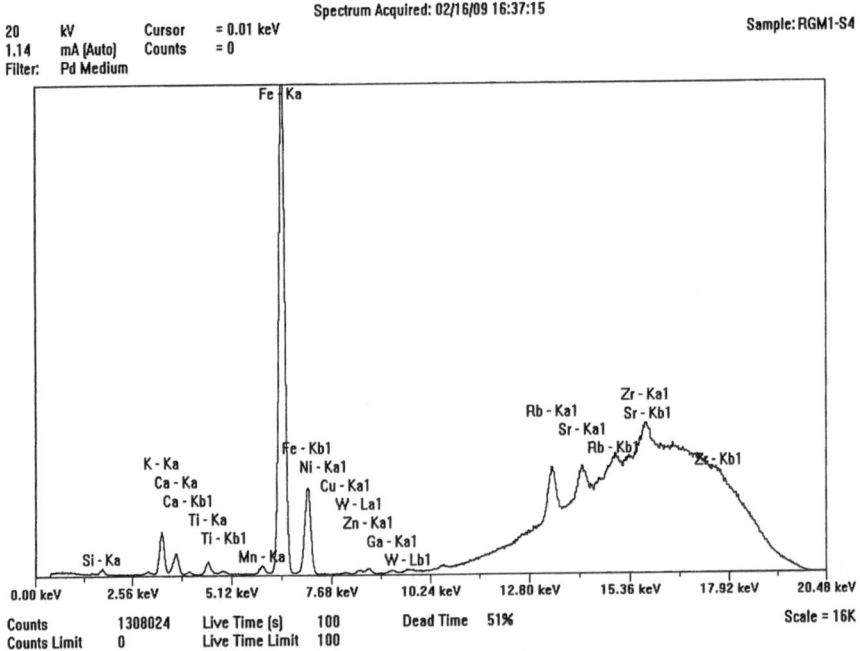

Fig. 2.8 A spectrum from a qualitative scan of RGM-1, showing the continuous radiation (bremsstrahlung) scatter apparent under the Rb–Zr peaks

well adapted. We use it frequently in the XRF lab at Berkeley just to determine the presence of elements in various solids, from unknown rocks to the relative proportions of elements on gold rings, or to determine whether a piece of jewellery is actually made of the metal the seller thinks it is made of.

Wavelength vs. EDXRF

Most of the discussions above have focused on EDXRF, and the descriptions of the instruments have been directed toward EDXRF. Wavelength XRF was the first to be used in archaeological analyses simply because it was the first invented, but by the 1970s EDXRF began to replace some of the WXRF systems in geological departments until the ICP-MS "revolution" in the 1990s. Today, WXRF is still the XRF system preferred in many geology institutions when an order of precision greater than EDXRF is required, particularly for light elements. Figures 2.9 and 2.10 show schematic renderings of both systems.

As in energy-dispersive systems, the WXRF X-ray source is quite similar. An anode of an appropriate element is paired with a target of an appropriate element for the elements of interest. However, in many modern WXRF systems, a number of tubes, and/or multiple targets are used in order to maximize the instrumental precision for $Z = 11$–92. In wavelength dispersive spectrometers, fluorescence

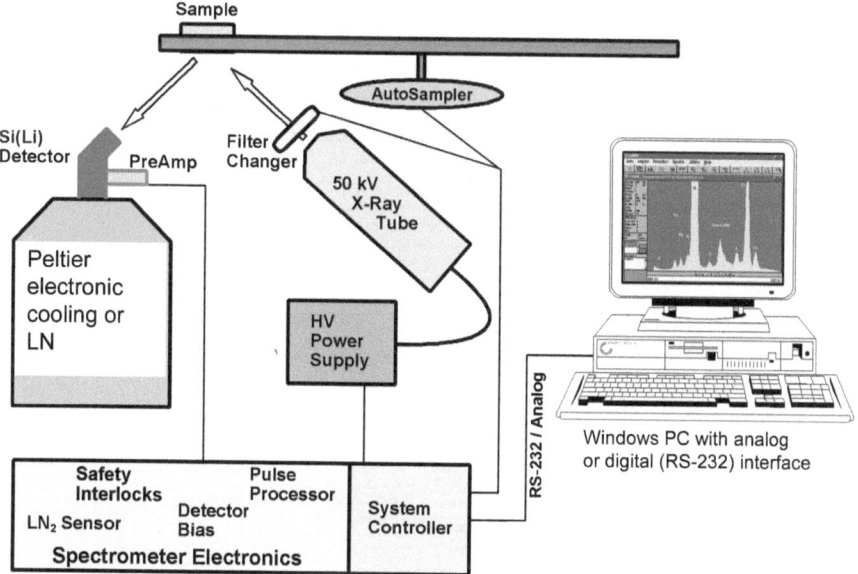

Fig. 2.9 Schematic drawing of a typical EDXRF instrument. Cooling can be either electronic Peltier or liquid nitrogen (LN)

X-ray photons are separated by *diffraction* on a single crystal before being detected. Although wavelength dispersive spectrometers are occasionally used to scan a wide range of wavelengths, producing a spectrum plot as in EDXRF, they are usually set up to make measurements only at the wavelength of the emission lines of the elements of interest, one element at a time. Collimation becomes even more important in WXRF to focus the incident X-rays toward the analysing crystal at the appropriated Bragg angle (Fig. 2.10). Importantly, unlike EDXRF, each element is selected by appropriate crystals individually according to Bragg's Law:

$$n \lambda = 2d \cdot \sin\theta$$
$$\text{or alternatively} \qquad (2.2)$$
$$\lambda = (2d/n)\sin\theta$$

where

- n is an integer determined by the order given
- λ is the wavelength of X-rays, and moving electrons, protons and neutrons
- d is the spacing between the planes in the atomic lattice, and
- θ is the angle between the incident ray and the scattering planes (see Fig. 2.9)

While the "early" WXRF instruments like the old Philips instruments at Berkeley and Arizona State University required manual selection of the analysing crystals with a manual goniometer, modern systems like the Philips 2400 WXRF at Berkeley are completely automated. After selecting the elements of interest and calibrating the instrument, the software selects the appropriate target(s) and crystal (s) based on that selection (see Jenkins, 1999 for more detail).

While instrumental precision in EDXRF systems has increased greatly in the last decade, WXRF instruments, simply because they are not separating all the seemingly chaotic energy entering the detector into individual channels and then into elemental data through preamplification, but selecting each element individually,

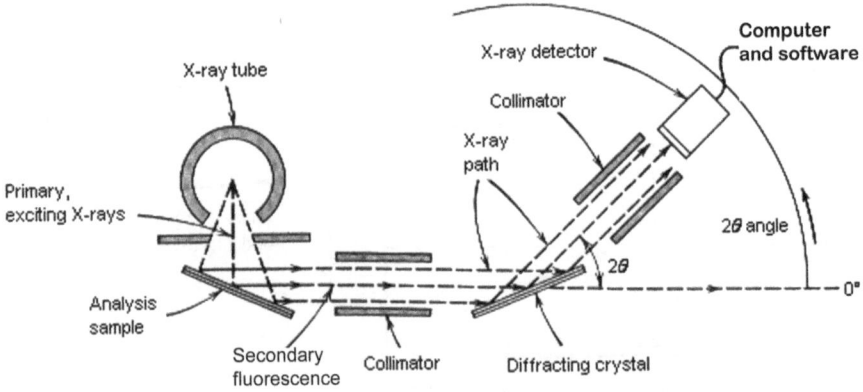

Fig. 2.10 Schematic drawing of a typical WXRF system. Only a single tube/target is shown here

Table 2.3 Detection limits (% for oxides, ppm for elements) for selected major and trace elements in whole rock EDXRF vs. WXRF (from Jenkins, 1999, 119)

Element	EDXRF	WXRF
Na_2O	0.81	0.16
Ti	0.008	0.006
Mn	0.002	0.014
Rb	3.0	0.6
Sr	2.8	0.4
Y	3.8	0.4
Zr	2.8	1.1
Nb	2.8	1.3

are much more precise particularly in the lighter elements or calculated oxides such as Na. Table 2.3 shows a comparison or instrumental precision between EDXRF and WXRF for various elements. Again, as I have said repeatedly, the analyst and/ or archaeologist must decide which level of precision and accuracy is desired or necessary (Shackley, 2002).

Non-Destructive WXRF Analyses

De Francesco et al. in Chap. 5 discuss a non-destructive qualitative analysis of archaeological obsidian with WXRF comparing favourably to destructive quantitative analyses. One of the major reasons that qualitative analyses have been favoured for WXRF work is that the Bragg angle assumes that the sample surface is absolutely flat or parallel to the crystal. Unfortunately, artifacts are not flat. So, by ratioing the elements against each other, the errors in the Bragg become "balanced" as I discussed above. However, most modern WXRF software will allow an analyst to perform a calibration based on ratioing to the Compton scatter just as in most EDXRF analyses. Tim Teague of the Department of Earth and Planetary Sciences at Berkeley and I have been doing this on the Philips PW2400 for a decade with precision equal to any EDXRF analysis, indeed without some of the issues with the heavier elements such as Ba in EDXRF (Shackley 1998a, 2005). The Philips *SuperQ* software has this utility, but few use it. One advantage of this is that a large number of elements can be acquired more rapidly than with the EDXRF systems that generally require a different set of conditions for each set of elements and instrument settings. The major limitation with most of the WXRF systems, as mentioned by De Francesco et al. in Chap. 5, is that the sample holders are of a limited size, 41 mm in the Philips PW 2400 case at Berkeley. In many EDXRF systems, the chamber is quite large and can be made even larger, as in the Hawaii-Hilo QuanX (Lundblad et al., 2008; Chap. 4). Additionally, the WXRF systems must run under vacuum. Air path analyses are not readily possible. So, while WXRF instruments can provide greater instrumental precision, there are limiting factors that can make EDXRF a better choice.

Analytical Instrument Settings and Providing Data

As part of the goal of this volume, we are trying to make XRF as understandable and transparent as we can to the curious archaeological community. We are also, of course, responsible for providing the data from our research to the public, which provides the bulk of our funding. For a number of years Mike Glascock (the author of Chap. 8) and I have been attempting to put all our obsidian source data on the web for the public, our supporting public. The elemental, geographic, and geological data from all known obsidian sources in the North American Southwest are on the web at: http://swxrflab.net/swobsrcs.htm and every effort is made to keep it up to date.

Immediately after a newly discovered source is published in print, I put the data online. Our job in the XRF community is to publish that data as soon as possible. Results are published in journals like *American and Latin American Antiquity, Antiquity, Archaeometry, Geoarchaeology, Journal of Archaeological Science*, and many other regional journals around the world.

Just as important is the publishing of the instrumental settings and analytical strategies used. This is necessary to evaluate one another's work, and for the archaeologist to understand what he or she is doing and whether the analytical strategy is sensible. I hope that after reading this chapter, the analytical trajectory and instrument settings that I give to you (the archaeologist) in each full report and in the Appendix here will be more comprehensible.

References

Asaro, F. and D. Adan-Bayewitz, (2007), The history of the Lawrence Berkeley National Laboratory Instrumental Neutron Activation Analysis Programme for Archaeological and Geological Materials. *Archaeometry* 49, 201–214.

Bertin, E., (1978), *Introduction to X-ray spectrometric analysis*. New York: Plenum.

Bouey, P., (1991), Recognizing the limits of archaeological applications of non-destructive energy-dispersive X-ray fluorescence analysis of obsidians. *Materials Research Society Proceedings* 185, 309–320.

Buras, B., Olsen, J.S., Andersen, A.L., Gerward, L., and Selsmark, B., (1974), Evidence of escape peaks caused by a Si(Li) detector in energy-dispersive diffraction spectra. *Journal of Applied Crystallography* 7:296–297.

Cann, J.R., (1983), Petrology of obsidian artifacts. In Kempe D.R.C., and Harvey A.P., (Eds.), *The Petrology of Archaeological Artefacts*, (pp. 227–255). Oxford: Clarendon.

Cann, J.R., and Renfrew, A.C.,(1964), The characterization of obsidian and its application to the Mediterranean region. *Proceedings of the Prehistoric Society* 30, 111–133.

Cesareo, R., Ridolfi, S., Marabelli, M., Castellano, A., Buccolieri, G., Donativi, M., Gigante, G.E., Brunetti, A., and Medina, M.A.R., (2008), Portable systems for energy-dispersive X-ray fluorescence analysis of works of art. In Potts, P.J., and West, M. (Eds.), *Portable X-ray fluorescence spectrometry: capabilities for in situ analysis*, (pp. 206–246). Cambridge: The Royal Society of Chemistry.

Craig, N., Speakman, R.J., Popelka-Filcoff, R.S., Glascock, M.D., Robertson, J.D., Shackley, M.S., Aldenderfer, M.S., (2007), Comparison of XRF and PXRF for analysis of archaeological obsidian from southern Perú. *Journal of Archaeological Science* 34 (12), 2012–2024.

Davis, M. K. 1994 Bremsstrahlung ratio technique applied to the non-destructive energy-dispersive X-ray fluorescence analysis of obsidian. *International association for obsidian studies bulletin* 11.

Davis, M.K., Jackson, T.L., Shackley, M.S., Teague, T., and Hampel, J., (1998), Factors affecting the energy-dispersive X-ray fluorescence (EDXRF) analysis of archaeological obsidian. In M.S. Shackley (Ed.), *Archaeological obsidian studies: method and theory*, (pp. 159–180). Advances in archaeological and museum studies 3. New York: Springer/Plenum Press.

Dillian, C.D., (2002), *More than toolstone: differential utilization of glass mountain obsidian*. Ph. D. dissertation, Department of Anthropology, University of California, Berkeley.

Eerkens, J.W., Ferguson, J.R., Glascock, M.D., Skinner, C.E., and Waechter, S.A., (2007), Reduction strategies and geochemical characterization of lithic assemblages: a comparison of three case studies from Western North America. *American Antiquity*, 72, 585–597.

Franzini, M., L. L. and Saitta M., (1976), Determination of the X-ray fluorescence mass absorption coefficient by measurement of the intensity of Ag Kα compton scattered radiation. *X-ray Spectrometry* 5, 84–87.

Giaque, R. D., Asaro, F., and Stross, F. H., (1993), High precision non-destructive X-ray fluorescence method applicable to establishing the provenance of obsidian artifacts. *X-ray Spectrometry*, 22, 44–53.

Glascock, M.D., Braswell, G.E., and Cobean, R.H., (1998), A systematic approach to obsidian source characterization. In M.S. Shackley (Ed.), *Archaeological obsidian studies: method and theory*, (pp. 15–66). Advances in archaeological and museum studies 3. New York: Springer/Plenum Press.

Goldberg, P., (2008), Raising the bar: making geological and archaeological data more meaningful for understanding the archaeological record. In Sullivan, A., Ed., *Archaeological concepts for the study of the cultural past*, (pp. 24–39). Salt Lake City: University of Utah Press.

Govindaraju, K., (1994), 1994 compilation of working values and sample description for 383 geostandards. *Geostandards Newsletter* 18 (special issue).

Hall, E.T., (1960), X-ray fluorescent analysis applied to archaeology. *Archaeometry* 3, 29–37.

Hampel, J.H., (1984), Technical considerations in X-ray fluorescence analysis of obsidian. In Hughes, R. E. (Ed.), *Obsidian Studies in the Great Basin*, (pp. 21–25). Berkeley: Contributions of the University of California Archaeological Research Facility 45.

Huckell, B.B., Holliday, V.T., Hamilton, M., Sinkovec, C., Merriman, C., Shackley, M.S., and R.H. Weber, The Mockingbird Gap Clovis Site: 2007 investigations. *Current Research in the Pleistocene* 25, 95–97.

Hughes, R.E., (1983), *Exploring diachronic variability in obsidian procurement patterns in northeast California and southcentral Oregon: geochemical characterization of obsidian-sources and projectile points by energy-dispersive X-ray fluorescence*. Ph.D. dissertation, Department of Anthropology, University of California, Davis.

Hughes, R.E., Ed., (1984). *Obsidian Studies in the Great Basin*. Berkeley: Contributions of the University of California Archaeological Research Facility 45.

Hughes, R.E., (2010) *Determining the geologic provenance of tiny obsidian flakes in archaeology using nondestructive EDXRF*. American Laboratory 42, 27–31.

Hull, Kathleen L., (2002) *Culture contact in context: a multiscalar view of catastrophic depopulation and culture change in Yosemite Valley*. Ph.D. dissertation, Department of Anthropology, University of California, Berkeley, CA.

Ivanenko, V.V., Kustov, V.N., and Matelev, A.Yu., (2003), Patterns of inter-elemental effects in EDXRF and a new correction method. *X-ray spectrometry* 32, 52–56.

Jack, R.N., (1971), The source of obsidian artifacts in northern Arizona. *Plateau* 43, 103–114.

Jack, R.N., (1976), Prehistoric obsidian in California: geochemical aspects. In Taylor, R.E., Ed., *Advances in obsidian glass studies: archaeological and geochemical perspectives*, (pp. 183–217). Park Ridge, NJ: Noyes Press.

Jack, R.N., and Carmichael, I.S.E., (1969), The chemical fingerprinting of acid volcanic rocks. *California division of mines and geology special report* 100, 17–32.

Jack, R.N. and Heizer, R.F., (1968) "Finger-Printing" of some Mesoamerican obsidian artifacts. Berkeley: Contributions of the University of California Archaeological Research Facility 5, 81–100.

Jackson, T.L., (1974), *The economics of obsidian in central California prehistory: applications of X-ray fluorescence spectrography in archaeology*. Master's thesis, Department of Anthropology, San Francisco State University, San Francisco.

Jackson, T.L., (1986), *Late prehistoric obsidian exchange in central California*. Ph.D. dissertation, Department of Anthropology, Stanford University.

Jackson, T.L. and Hampel, J.H. (1992). Size Effects in the Energy-Dispersive X-ray Fluorescence (EDXRF) Analysis of Archaeological Obsidian Artifacts. Poster presented at the 28th International Symposium on Archaeometry, Los Angeles.

Jenkins, R., (1999), *X-ray fluorescence spectrometry: second edition*. New York: Wiley-Interscience.

Jenkins, R., Gould, R.W., and Gedcke, D., (1981), *Quantitative X-ray spectrometry*. New York: Marcel Dekker.

Joyce, R.A., Shackley, M.S., Sheptak, R. and McCandless, K., (2004), Resultados preliminares de una investigación con EDXRF de obsidiana de Puerto Escondido. In *memoria, vii seminario de antropología de Honduras "Dr. George Hasemann"*, (pp. 115–130). Instituto Hondureño de Antropología e Historia, Honduras.

Kahn, J.G., (2005) *Household and community organization in the late prehistoric Society Island Chiefdoms (French Polynesia)*. Ph.D. dissertation, Department of Anthropology, University of California, Berkeley.

Killick, D., (2008), Archaeological science in the USA and in Britain. In Sullivan, A., Ed., *Archaeological concepts for the study of the cultural past*, (pp. 40–64). Salt Lake City: University of Utah Press.

Lachance, G.R., and Claisse, F., (1994), Quantitative X-ray fluorescence analysis. New York: Wiley-Interscience.

Lundblad, S. P., Mills, P. R., & Hon, K. (2008). Analysing archaeological basalt using non-destructive energy-dispersive X-ray fluorescence (EDXRF): Effects of post-depositional chemical weathering and sample size on analytical precision. *Archaeometry, 50*, 1–11.

Martynec, R., Davis, R., and Shackley, M.S., (2010), The Los Sitios del Agua Obsidian source (formerly AZ unknown a) and recent archaeological investigations along the Rio Sonoyta, northern Sonora. *Kiva*, in press.

McCarthy, J.J., and Schamber, F.H. (1981) Least-squares fit with digital filter: a status report. In Heinrich, K.F.J., Newbury, D.E., Myklebust, R.L., and Fiori, E. (Eds.), *Energy Dispersive X-ray Spectrometry*, (pp. 273–296). Washington, D.C.: National Bureau of Standards Special Publication 604.

Moseley, H.G.J., (1913/1914), High frequency spectra of the elements. *The philosophers magazine*, 26, 1024–1034, and 27, 703–713.

Nazaroff, A., and Shackley, M.S., (2009), Testing the size dimension limitation of portable XRF instrumentation for obsidian provenance. Poster presentation, Geological Society of America Annual Meeting, Portland, OR.

Negash, A. and Shackley, M.S., (2006), Geochemical provenance of obsidian artefacts from the MSA site of Porc Epic, Ethiopia. *Archaeometry* 48, 1–12.

Negash, A., Shackley, M.S. and Alene, M., (2006) Source provenance of obsidian artifacts from the Early Stone Age (ESA) site of Melka Konture, Ethiopia. *Journal of Archaeological Science* 33, 1647–1650.

Pessanha, S., Guilherme, A., and Carvalho, M.L., (2009), Comparison of matrix effects on portable and stationary XRF spectrometers for cultural heritage samples. *Applied Physics A* 97, 497–505.

Phillips, S.C., and Speakman, R.J., (2009), Initial source evaluation of archaeological obsidian from the Kuril Islands of the Russian Far East using portable XRF. *Journal of Archaeological Science* 36, 1256–1263.

Potts, P.J. and West, M., Eds., (2008), *Portable X-ray fluorescence spectrometry: capabilities for in situ analysis.* Cambridge: The Royal Society of Chemistry.

Rindby, A., (1989), Software for energy-dispersive X-ray fluorescence. *X-ray spectrometry*, 18, 113–118.

Röntgen, W.K., (1898), On a new kind of rays: second communication. *Annals of Physical Chemistry*, 64, 1–11.

Schamber, F.H., (1977) A modification of the linear least-squares fitting method which provides continuum suppression. In Dzubay, T.G. (Ed.), *X-ray Fluorescence Analysis of Environmental Samples*, (pp. 241–257). Ann Arbor: Ann Arbor Science.

Shackley, M. S., (1988), Sources of archaeological obsidian in the Southwest: an archaeological, petrological, and geochemical study. *American Antiquity* 53, 752–772.

Shackley, M.S., (1990). *Early hunter-gatherer procurement ranges in the Southwest: evidence from obsidian geochemistry and lithic technology.* Ph.D. dissertation. Tempe: Arizona State University.

Shackley, M.S., (1991) Tank Mountains obsidian: a newly discovered archaeological obsidian source in east-central Yuma County, Arizona. *Kiva* 57, 17–25.

Shackley, M.S., (1992). The Upper Gila River gravels as an archaeological obsidian source region: implications for models of exchange and interaction. *Geoarchaeology* 4, 315–326.

Shackley, M.S., (1995), Sources of archaeological obsidian in the greater American Southwest: an update and quantitative analysis. *American Antiquity* 60, 531–551.

Shackley, M.S., (1998a), Geochemical differentiation and prehistoric procurement of obsidian in the Mount Taylor Volcanic Field, Northwest New Mexico. *Journal of Archaeological Science* 25, 1073–1082

Shackley, M.S., (1998b) Intrasource chemical variability and secondary depositional processes in sources of archaeological obsidian: lessons from the American Southwest. In Shackley, M.S. (Ed.), *Archaeological obsidian studies: method and theory* (pp. 83–102). Advances in archaeological and museum science 3. New York: Springer/Plenum.

Shackley, M.S., Ed., (1998c), *Archaeological obsidian studies: method and theory.* Advances in Archaeological and Museum Science 3, New York: Springer/Plenum Publishing Corporation.

Shackley, M.S., (2002) Precision versus Accuracy in the XRF analysis of archaeological obsidian: some lessons for archaeometry and archaeology. In Jerem, E., and Biro, K.T., Eds. *Proceedings of the 31st Symposium on Archaeometry, Budapest, Hungary,* (pp. 805–810). Oxford: British Archaeological Reports International Series 1043 (II).

Shackley, M.S., (2005), *Obsidian: geology and archaeology in the North American Southwest.* Tucson: University of Arizona Press.

Shackley, M.S., (2010), Is there reliability and validity in portable X-ray fluorescence spectrometry (PXRF)? SAA Archaeological Record (in press).

Shackley, M.S. and Dillian, C., (2002), Thermal and environmental effects on obsidian geochemistry: experimental and archaeological evidence. In Loyd, J.M, Origer, T. M. and Fredrickson, D.A. (Eds.), *The effects of fire and heat on obsidian,* (pp. 117–134). Sacramento: Cultural resources publication, anthropology-fire history, U.S. Bureau of Land Management.

Shackley, M. S. and Hampel, J., (1992), Surface effects in the energy dispersive X-ray fluorescence (EDXRF) analysis of archaeological obsidian. Poster presented at the 28th International Symposium on Archaeometry, Los Angeles.

Silliman, S., (2000), *Colonial worlds, indigenous practices: the archaeology of labor on a 19th century California rancho.* Ph.D. dissertation, Department of Anthropology, University of California, Berkeley.

Speakman, R.J. and Neff, H., Eds., (2005), *Laser ablation ICP-MS in archaeological research*. Albuquerque: University of New Mexico Press

Weisler, M. I., (1993), *Long-distance interaction in Prehistoric Polynesia: three case studies*. Ph. D. dissertation, Department of Anthropology, University of California, Berkeley.

Weisler, M.I., and Woodhead, J.D., (1995), Basalt Pb isotope analysis and the prehistoric settlement of Polynesia. *Proceedings of the National Academy of Science*, 92, 1881–1885

Williams-Thorpe, O., (2008), The application of portable X-ray fluorescence analysis to archaeological lithic provenancing. In Potts, P.J., and West, M. (Eds.), *Portable X-ray fluorescence spectrometry: capabilities for in situ analysis*, (pp. 174–205). Cambridge: The Royal Society of Chemistry.

Chapter 3
Factors Affecting the Energy-Dispersive X-Ray Fluorescence (EDXRF) Analysis of Archaeological Obsidian

M. Kathleen Davis, Thomas L. Jackson, M. Steven Shackley, Timothy Teague, and Joachim H. Hampel

Introduction to the Springer Reprint

At the 28th International Symposium on Archaeometry in 1992 in Los Angeles, Joachim Hampel, Tom Jackson, and Steve Shackley, presented two poster presentations on size and surface effects in the energy-dispersive X-ray fluorescence (EDXRF) analysis of archaeological obsidian (Jackson and Hampel, 1992; Shackley and Hampel, 1992). Joachim Hampel, who was the XRF technician in what was then called the Department of Geology and Geophysics, and is now Earth and Planetary Sciences at Berkeley, was instrumental in worrying about these issues and was just as instrumental in shaping the authors of this 1998 paper, as well as many others from the 1970s through the 1990s till his retirement. In 1984, in Richard Hughes (1984) volume on Great Basin obsidian, his "Technical considerations in X-ray fluorescence analysis of obsidian" was the first major step toward understanding the real resolution of EDXRF for archaeological obsidian. Kathy Davis, the primary author and experimental genius who did the work and synthesis for this chapter, provided one of the most referenced papers on the subject in archaeology today. Indeed, the size and surface effects that are integrated into "normal" science in EDXRF studies are based on this project and paper. The Lundblad et al. paper presented here is very much based on the assumptions generated by this paper and replicated in basalt studies, as discussed by these authors. For this reason, we republish this important work more than 10 years later, but it is just as critical today. I have eliminated the discussion of EDXRF theory and instrumentation, including two figures, and refer the reader to Chap. 2, which is a much more intensive exploration of the subject with the archaeologist in mind. Minor editorial changes have been made to fit the publisher's guidelines. Otherwise, the text, tables, and illustrations are exactly as in the original.

M. Steven Shackley (✉)
Far Western Anthropological Research Group, 2727 Del Rio Place, Suite A, Davis,
CA 95618, USA
e-mail: shackley@berkeley.edu

M.S. Shackley (ed.), *X-Ray Fluorescence Spectrometry (XRF) in Geoarchaeology*,
DOI 10.1007/978-1-4419-6886-9_3, © Springer Science+Business Media, LLC 2011

Introduction

Non-destructive X-ray fluorescence (XRF) is now widely accepted by archaeologists as a tool for the identification of the geologic origins of obsidian artifact raw material, and to a lesser extent, artifacts of basalt and other volcanic rock. While the advantages of the non-destructive EDXRF approach are obvious for artifact studies, analysis of unmodified obsidian rather than powdered samples contributes to an already long list of analytical uncertainties inherent in XRF analysis. Understanding the magnitude and source of these uncertainties is crucial to accurate ascription of artifacts to source, particularly when characterizing artifacts from closely related sources. Previous attempts to quantify analytical uncertainties in non-destructive analysis have focused on errors related to variation in artifact size and surface morphology (Bouey, 1991; Jackson and Hampel, 1992; Shackley and Hampel, 1992; Davis, 1994). These studies suggest that artifact size is potentially the more important contributor although the actual magnitude of errors due to surface morphology remains poorly understood.

Elaborating on these earlier studies, this chapter attempts to quantify errors associated with artifact size and surface morphology via two different experiments. The first experiment is an analysis of unmodified samples of various sizes made from a single obsidian core; the second involves multiple analyses of flaked and powdered samples to measure errors related to artifact surface variability. All samples are analyzed for the elements Ti, Mn, Fe, Zn, Pb, Rb, Sr, Y, Zr, and Ba, using analyte/scatter and peak ratio techniques to compensate for errors associated with the analysis of unmodified obsidian. Since both experiments involve multiple analyses of samples and comparisons to powdered samples, it is also hoped that the data may be useful as a general indicator of the accuracy and precision of the method.

Methods

Size Experiment

To determine the limits of the scatter and peak ratio techniques as discussed in Chap. 2 and to correct for artifact size, we analyzed 20 samples of varying thickness and diameter prepared from a single core of obsidian from the Glass Mountain source in northeastern California. Samples were prepared at the Department of Geology and Geophysics, University of California at Berkeley, and are also the subject of a paper by Jackson and Hampel (1992). Samples were carefully cut and ground to specified dimensions using a lapidary saw and abrasive, and are smooth and polished on both target and opposite sides. The first eight samples are round, with a fixed diameter of 25 mm and thicknesses of 0.03, 0.2, 0.5, 1, 2, 3, 4, and 5 mm. By stacking various combinations of these eight samples, analyses were also conducted for sample thicknesses of 6–11 mm, and for the 1.2, 1.5, 1.7, and 2.5 mm thick samples reported in the mid-Z analysis (Table 3.2). The remaining 12 samples are square, measuring 3, 5, 10, or 15 mm on a side with thicknesses of 1, 2, and 3 mm for each.

Table 3.1 Selected trace element concentrations of two international rock standards

	Zn	Pb	Rb	Sr	Y	Zr	Ti	Mn	Fe	Ba	Ba (Am)
RGM-1											
Govindaraju 1989	32	24	149	108	25	219	1601	279	1.86	807	807
(this study)	44	20	149	107	25	218	1673	288	2	794	785
±	4.9	2.5	2.8	6.3	1.6	5.1	81.3	29.7	0.10	9.2	12.8
NBS-278											
Govindaraju 1989	55	16.4	127.5	63.5	41	295	1469	403	2.04	1140	1140
(this study)	53	16	128	65	40	284	1407	450	2	–	–
±	5.1	2.4	2.7	6.3	1.6	5.2	81.3	29.8	0.10	–	–

Plus or minus values represent a combined estimate of counting and fitting error uncertainties. All values are in ppm with the exception of Fe ($Fe_2O_3{}^T$), which is in weight percent. RGM-1 is a U.S. Geological Survey rhyolite (obsidian) standard, and NBS-278 is a National Bureau of Standards obsidian standard. Values for barium acquired with the [241]Am radioisotope source are labeled Ba(Am)

Surface Variability Experiment

A second experiment, using a different set of obsidian samples, was devised to measure errors resulting from the irregular surfaces of obsidian artifacts. These samples, also prepared at the Department of Geology and Geophysics at Berkeley, are the subject of a paper on surface effects by Shackley and Hampel (1992). With a fixed diameter of 25 mm, the target side of each sample has been flaked with a copper flaking tool to approximate the surface of a typical flaked stone artifact. Ten such samples are analyzed here, five from the Bodie Hills source located in eastern California and five from the Napa Valley source located in the North Coast Ranges of California. Each set of five samples, designated here BH-1 through BH-5 for Bodie Hills, and NV-1 through NV-5 for Napa Valley, was prepared from a single core of obsidian. The Bodie Hills samples range from 5 to 7 mm thick, while the Napa Valley samples range from 8 to 13 mm thick.

The flaked surface of each sample was analyzed in a random orientation and then rotated 45, 90, 180, 270, and 315° from the original orientation for five subsequent analyses. These results were then compared to analyses of pressed powder samples, one each of Bodie Hills and Napa Valley obsidian, which were acquired in the same six orientations. The Bodie Hills and Napa Valley powdered samples were not prepared from the same core as the flaked samples, and thus are not strictly comparable. However, our primary concern here is the comparison of relative and not absolute errors. Bodie Hills and Napa Valley powdered samples were chosen so that they would have peak counts comparable to those of the flaked samples. Two additional powdered samples, U.S. Geological Survey (USGS) standard RGM-1

Table 3.2 Trace element values for unmodified (i.e., unpowdered) Glass Mountain obsidian of varying diameter and thickness

Diameter (mm)	Thickness (mm)	Zn	Pb	Rb	Sr	Y	Zr	Fe/Mn	Fe/Ti	Ti	Mn	Fe	Ba	Ba(Am)
Control group														
25	11	38	20	152	112	29	233	67	35	1576	281	1.87	781	533
25	10	41	21	156	111	29	233	69	36	1560	270	1.87	813	538
25	9	38	25	155	112	28	234	63	34	1620	297	1.87	794	539
25	8	38	21	155	115	28	236	67	35	1605	279	1.87	824	553
25	7	43	23	161	112	25	229	68	34	1630	274	1.88	778	559
25	6	39	24	154	111	27	235	65	36	1545	286	1.86	798	564
25	5	34	20	158	112	27	230	71	35	1635	268	1.91	767	597
25	4	35	24	154	114	27	231	75	35	1613	256	1.93	814	640
	Mean	38	22	156	112	28	233	68	35	1598	276	1.9	796	565
	SD	2.8	1.9	2.6	1.5	1.2	2.5	3.7	0.6	33.4	12.3	0.03	20.2	36.4
	CV	7.4	8.6	1.7	1.4	4.4	1.1	5.5	1.8	2.1	4.5	1.4	2.5	6.4
Fixed diameter														
25	3	**47**	23	158	115	27	232	73	34	1633	256	1.90	793	642
25	2.5	36	**15**	154	115	25	235							
25	2	44	22	157	**117**	30	234	75	34	1624	**247**	1.87	815	**670**
25	1.7	**45**	20	**161**	116	26	230							
25	1.5	**55**	21	161	**119**	27	235							
25	1.2	**50**	19	**172**	**123**	**31**	238							
25	1	**47**	24	**184**	**130**	28	**250**	74	35	1597	252	1.87	784	**730**
25	0.5	**61**	**32**	**207**	**145**	29	**270**	71	34	1638	260	1.87	755	**744**
25	0.2	**86**	**45**	**251**	**167**	**35**	**285**	74	34	1630	254	1.89	784	**770**
25	0.03	**185**	**60**	**240**	**159**	29	**256**	65	**30**	**1416**	**223**	**1.43**	**602**	**1177**
Fixed thickness														
25	3	**47**	23	158	115	27	232	73	34	1633	256	1.90	793	642
15	3	36	20	155	115	27	234	75	36	1536	**245**	1.84	**852**	592
10	3	**59**	22	151	110	**25**	231	71	36	**994**	**164**	**1.14**	824	**476**
5	3	**56**	18	**135**	98	27	**205**	71	35	**275**	**38**	**0.18**	656	**221**
3	3	45	**11**	**102**	88	**20**	180	55	35	**130**	18	**0.00**	418	**93**
25	2	44	22	157	**117**	30	234	75	34	1624	247	1.87	815	**670**
15	2	**51**	19	151	114	**31**	228	76	36	1604	254	1.93	**845**	**659**
10	2	42	20	161	115	27	**239**	74	34	**987**	**150**	**1.08**	780	**417**
5	2	**69**	**15**	**135**	95	22	**201**	59	**28**	**339**	**48**	**0.20**	644	**158**
3	2	**49**	**16**	**108**	74	17	**162**	91	34	**139**	12	**0.01**	410	**64**
25	1	**47**	24	**184**	**130**	28	**250**	74	35	1597	252	1.87	784	**730**
15	1	37	26	**178**	**128**	26	**248**	70	35	**1399**	**235**	**1.65**	769	**477**
10	1	**47**	17	**175**	**124**	27	**245**	71	34	**953**	**152**	**1.04**	805	**360**
5	1	**65**	26	**133**	91	21	**194**	75	29	**332**	**38**	**0.20**	568	**111**
3	1	41	**14**	**100**	76	**24**	**160**	97	34	**131**	**10**	**-0.00**	274	**34**

All element values are expressed in ppm, with the exception of Fe/Mn and Fe/Ti peak ratios, and Fe (Fe$_2$O$_3$^{T}), which is expressed in weight percent. Values for barium acquired with the [241]Am radioisotope source are labeled Ba(Am). Values in the lower half of the table appear in bold type when they differ significantly (two sample *t*-test with $\alpha = 0.05$) from the control group above and Geological Survey of Japan standard JR-2, were also included in the analysis and analyzed in the six orientations described above.

Analytical Conditions

Both sample sets analyzed in this study were subjected to the following four procedures. Analyses for the elements Ti, Mn, Fe, Zn, Pb, Rb, Sr, Y, Zr, and Ba were conducted on a Spectrace 5000 EDXRF spectrometer, at BioSystems Analysis Inc. The Spectrace 5000 is equipped with a Si(Li) detector with a resolution of 155 eV FWHM for 5.9 keV X-rays (at 1,000 counts per second) in an area 30 mm^2. Signals from the spectrometer are amplified and filtered by a time variant pulse processor and sent to a 100 MHz Wilkinson type analog-to-digital converter. The X-ray tube employed is a bremsstrahlung type, with an Rh target and a 5 mil Be window. The tube is driven by 50 kV, 1 mA high voltage power supply, providing a voltage range of 4–50 kV (Fig. 3.1a)

For analysis of the mid-Z elements (Zn, Pb, Rb, Sr, Y, and Zr), the Rh X-ray tube is operated at 30 kV, 0.30 mA (pulsed), with a 0.127 mm Pd filter. Analytical lines used are Zn (K-a), Pb (L-a), Rb (K-a), Sr (K-a), Y (K-a), and Zr (K-a). Samples are scanned for 200 s live-time in an air path. Peak intensities for the above elements are calculated as ratios to the Compton scatter peak of rhodium and converted to parts-per-million (ppm) by weight using linear regressions derived from the analysis of 20 USGS, U.S. National Bureau of Standards (NBS) and Geological Survey

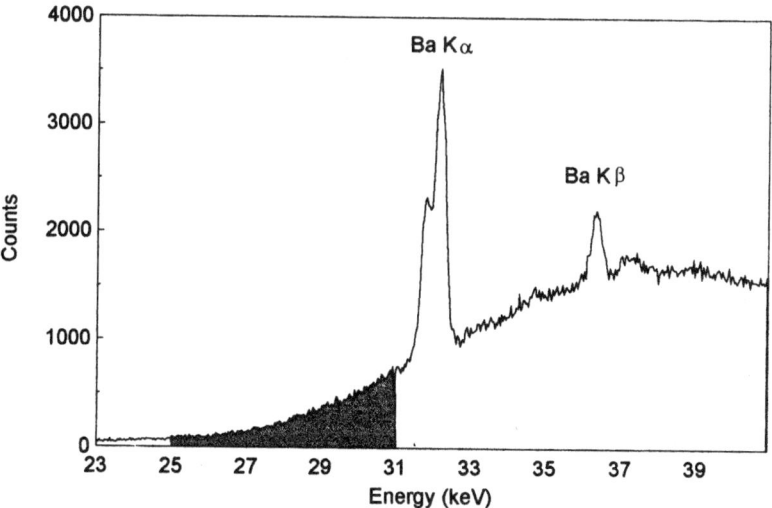

Fig. 3.1 X-ray spectrum of a sample of unmodified Glass Mountain obsidian acquired at 50 kV showing the K-a and K-b peaks of Ba. The darkened bremsstrahlung region between 25 and 31 keV is used to correct for sample mass and thickness

of Japan (GSJ) rock standards. The analyte/Compton scatter ratio is employed to correct for variation in sample size, surface morphology, and sample matrix (see Chap. 2) (Fig. 3.2).

For analysis of the elements Ti (TiO_2), Mn (MnO), and Fe ($Fe_2O_3^T$), the X-ray tube is operated at 12 kV, 0.27 mA with a 0.127 mm aluminum filter. Samples are scanned for 200 s live-time in a vacuum path. Element values are reported in two ways: in units of concentration (parts per million for Ti and Mn, and weight percent for total Fe); and as Fe/Mn and Fe/Ti peak ratios. Peak ratios are calculated as simple ratios of extracted peak intensities ($K\alpha$ and $K\beta$). Concentration values are calculated using linear regressions derived from the analysis of 13 standards from the USGS, the NBS and the GSJ. These values are reported to better evaluate the precision of each measurement. It should be noted that the concentration values are *not* corrected against a spectral reference, however, and are corrected for matrix effects only to the extent that the concentration range of Fe is limited in the chosen standards. As a result, large errors are evident among the smaller samples (Fig. 3.3).

For analysis of Ba, the X-ray tube is operated at 50 kV, 0.25 mA with a 0.63 mm Cu filter in the X-ray path. Samples are scanned for 200 s live-time in an air path. Trace element intensities for Ba, extracted from a net fit of the $K\alpha$ peak, are calculated as ratios to the Bremsstrahlung region between 25.0 and 30.98 keV (Fig. 3.1b). Ppm values are generated using a polynomial regression derived from the analysis of eight USGS and eight GSJ rock standards.

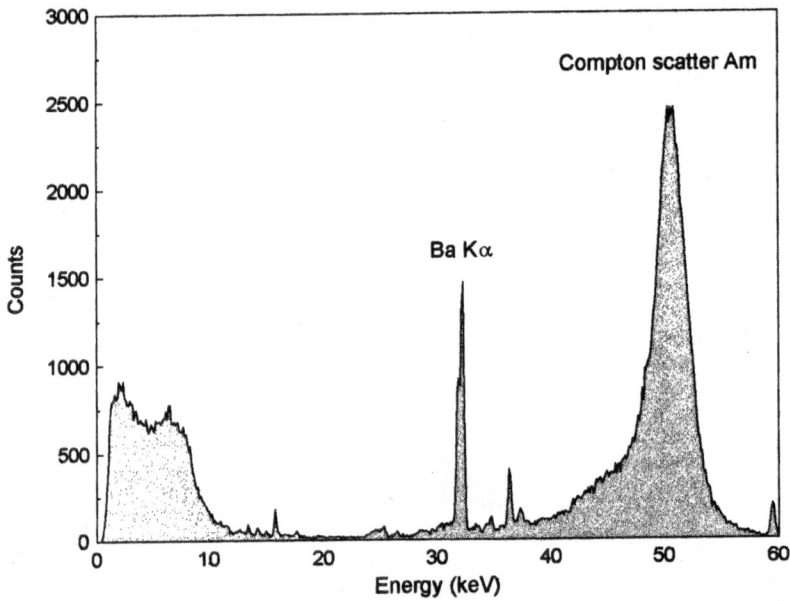

Fig. 3.2 X-ray spectrum of unmodified Glass Mountain obsidian acquired using an [241]Am radioisotope source. Labeled are the K-a peak of Ba and the Compton scatter peak Am

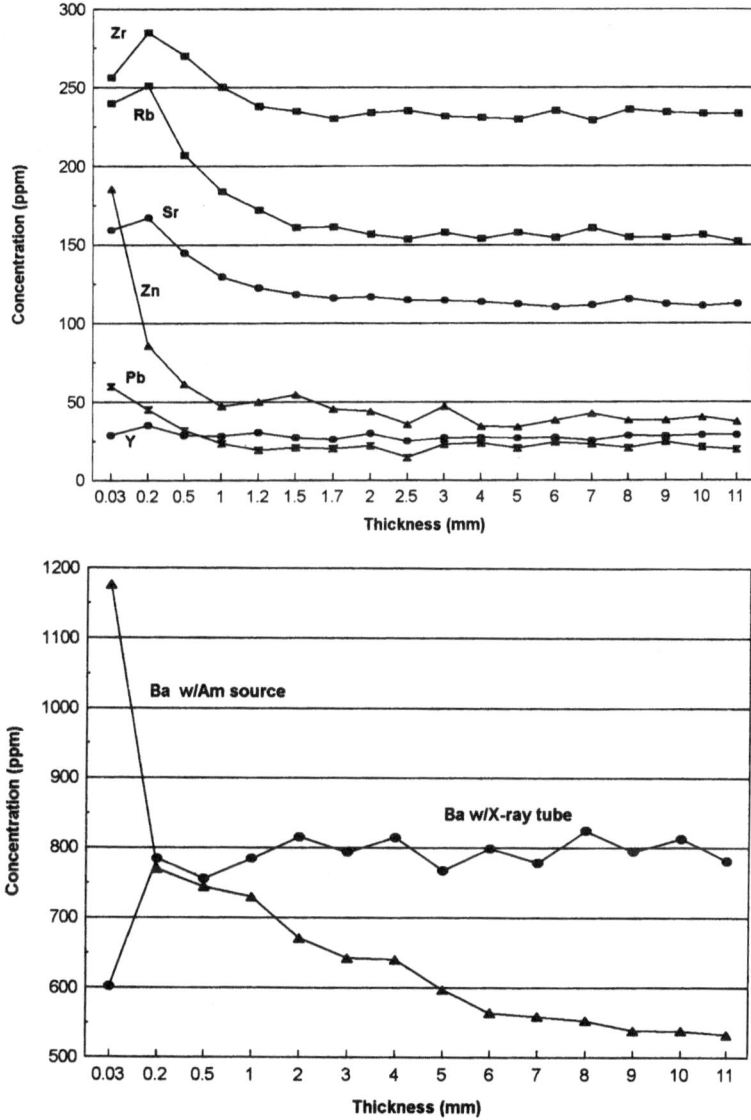

Fig. 3.3 (**a–d**) Trace element values for unmodified Glass Mountain obsidian as a function of sample thickness. All concentrations reported in ppm, with the exception of Fe (Fe$_2$O$_3^T$) which is reported in weight percent. All samples have a fixed diameter of 25 mm. (**a**) mid-Z elements; (**b**) Ba acquired at 50 kV with an X-ray tube, and Ba acquired with a ^{241}Am radioisotope source; (**c**) Ti, Mn, and Fe. Ti and Mn are in ppm (left *y-axis*), and Fe is in weight percent (right *y-axis*); (**d**) Fe/Ti and Fe/Mn peak ratios

Samples were analyzed a second time for Ba using a [241]Am radioisotope source on a Spectrace 440 at the XRF lab in the Department of Geology and Geophysics at the University of California, Berkeley. The system is equipped with a Si(Li) detector with a resolution of 142 eV FWHM at 5.9 keV in an area approximately 20 mm^2. Peak intensities for Ba were obtained by irradiating specimens with the radioisotope source for 200 s lifetime. Peak intensities were calculated as ratios to the Compton scatter peak of Am and converted to ppm by weight using linear regressions derived from the analysis of ten rock standards from the USGS, the NBS, and the GSJ.

With the exception of analysis for Ba on the Spectrace 5000 (see above), peak intensities for elements in both systems are extracted using the Super ML data analysis routine, where peak intensities for both Kα and Kβ peaks are extracted and corrected for overlap and background via comparison to stored reference standards (McCarthy and Schamber, 1981; Schamber, 1977). Analytical results for selected reference standards are given in Table 3.1 (Fig. 3.4).

Results

Size Experiment

Analytical results and sample dimensions for the Glass Mountain samples are reported in Table 3.2 and presented in Figs. 3.5 and 3.6. The accuracy of these results may be evaluated against reported element values for RGM-1, a USGS rhyolite standard also derived from the Glass Mountain obsidian flow (Tatlock et al., 1976). Trace element values for RGM-1 reported by Govindaraju (1989) are given in Table 3.1. In general, agreement between the larger Glass Mountain specimens (i.e., samples thicker than 3 mm and with a diameter of 25 mm) and RGM-1 values is quite good for most elements, though some systematic errors are evident. In the mid-Z analysis, element values for the elements with adequate counting statistics (i.e., Rb, Sr, and Zr) are, on average, 5% higher than the values reported for RGM-1. We suspect that this error is related to the calculation of element values for unmodified obsidian samples against calibration lines derived from the analysis of powders, due to the lower element/Compton ratio of powdered samples compared with that of unprocessed obsidian samples. Analyses of Ba with the radioisotope source are exceptionally poor. In this case, only 3 of the 26 measurements made fall within 10% of the 807 ppm reported by Govindaraju. It appears that the analyte/Compton ratio is not correcting for sample size, or that some other source of error is present.

To evaluate the precision of the results, and thus determine the point at which the trace element values are measurably affected by sample size, we compare element values of the smaller specimens to values of those samples that are well above the estimated size limits. For this purpose, we have chosen samples with a diameter of 25 mm and a minimum thickness of 4 mm as a control group. The 25 mm diameter

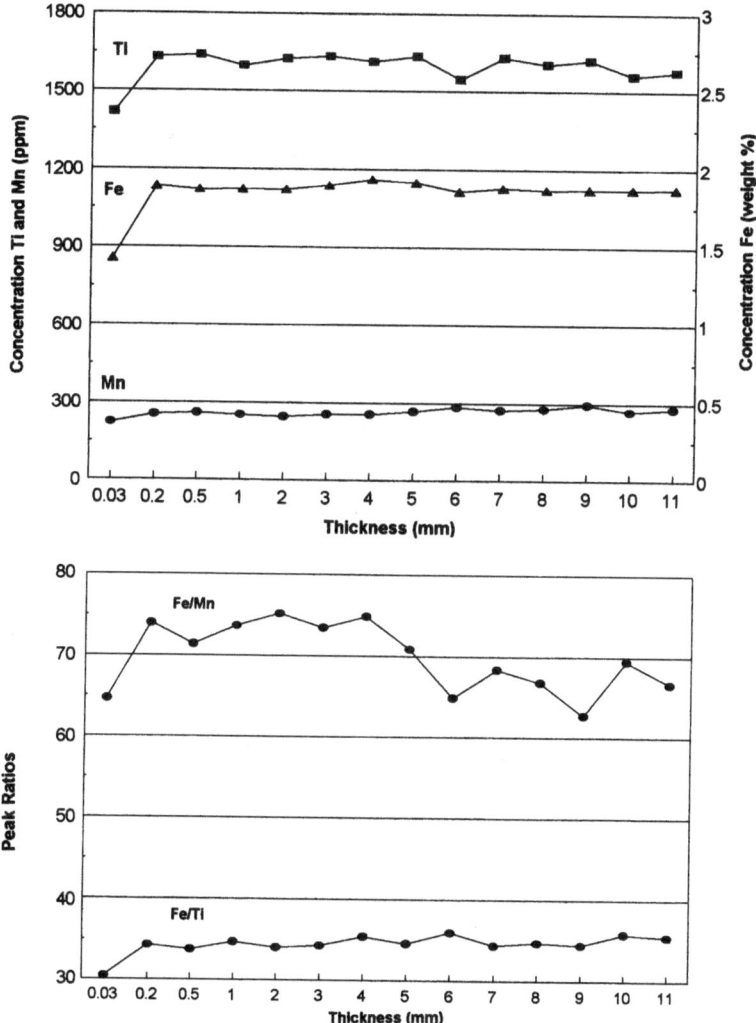

Fig. 3.4 (**a–d**) Trace element values for unmodified Glass Mountain obsidian as a function of sample diameter. All concentrations reported in ppm, with the exception of Fe ($Fe_2O_3^T$) which is reported in weight percent. Samples are square with a fixed thickness of 3 mm. (**a**) mid-Z elements; (**b**) Ba acquired at 50 kV with an X-ray tube, and Ba acquired with a [241]Am radioisotope source; (**c**) Ti, Mn, and Fe. Ti and Mn are in ppm (left y-*axis*), and Fe is in weight percent (right y-*axis*); (**d**) Fe/Ti and Fe/Mn peak ratios

samples completely cover the sample slot so that sample diameter or skewed placement of a sample should not factor into the measurements. The minimum thickness of 4 mm in the control group is well above infinite thickness for all elements except Ba, for which infinite thickness is over a centimeter at this energy. This is not ideal. However, a test of all element values in the control group revealed no systematic variation in element values relative to thickness, except for Ba

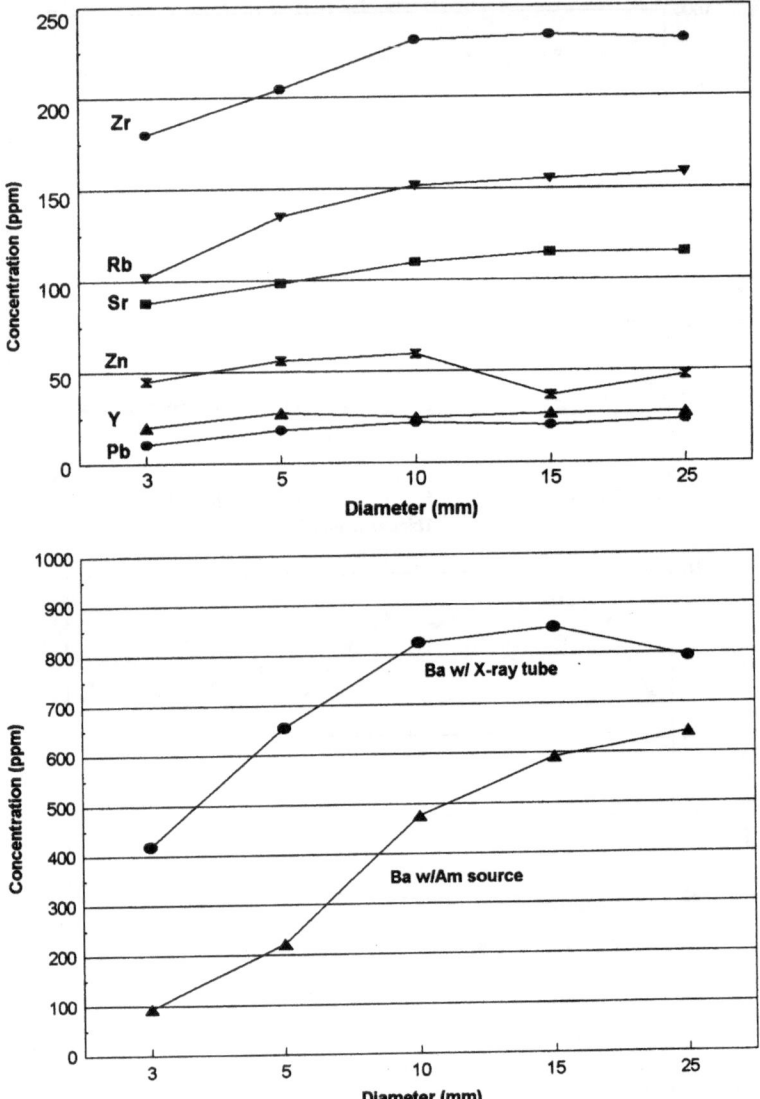

Fig. 3.5 Relative errors acquired from six runs of Bodie Hills obsidian. Each point represents the coefficient of variance of a given element in a single sample analyzed in six different orientations. Squares represent CV for unmodified samples of obsidian with flaked surfaces (BH-1 through BH-5), while the triangle represents the CV of a pressed powder sample of Bodie Hills obsidian. Numbers in bold above each element indicate the average concentration of that element. All values are in ppm with the exception of Fe/Ti and Fe/Mn peak ratios, and Fe ($Fe_2O_3^T$) which is in weight percent. Values for Ba acquired with the [241]Am radioisotope source are labeled Ba/Am

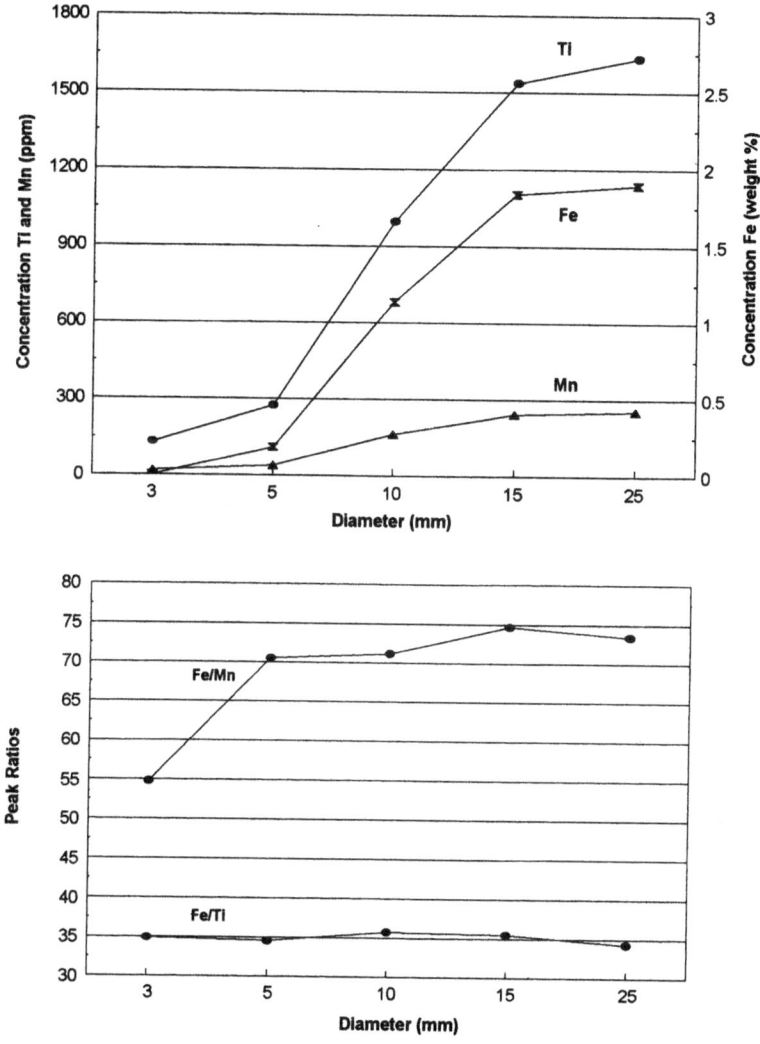

Fig. 3.6 Relative errors acquired from six runs of Napa Valley obsidian. Each point represents the coefficient of variance of a given element in a single sample analyzed in six different orientations. Squares represent CV for unmodified samples of obsidian with flaked surfaces (NV-1 through NV-5), while the triangle represents the CV of a pressed powder sample of Napa Valley obsidian. Numbers in *bold* above each element indicate the average concentration of that element. All values are in ppm with the exception of Fe/Ti and Fe/Mn peak ratios, and Fe ($Fe_2O_3^T$) which is in weight percent. Values for Ba acquired with the [241]Am radioisotope source are labeled Ba/Am

acquired with the radioisotope source. As mentioned above, precision and accuracy are poor in this analysis for all samples. For all other elements, significant deviation from the control group averages in the smaller samples should indicate the point at which sample size has affected the measurement. In Table 3.2, element values for the smaller samples appear in bold type when they differ significantly (two-sample *t*-test at $a = 0.05$) from the control sample above. To better emphasize the general trends in the data, selected element values are depicted as a function of sample thickness in Fig. 3.5a–d and as a function of sample diameter in Fig. 3.6a.d.

At the physical level, observed element distortions are the result of two conditions – infinite thickness and the field of view of the detector. When a sample is too small to cover the area seen by the detector, the relative deficit in X-rays from the sample results in element intensities that are lower than one would expect for a given material. This effect is best illustrated in the Ti, Mn, and Fe concentration values plotted in Fig. 3.6c, where no ratio correction is employed. In this case, element values fall drastically at the 10 mm diameter.

As discussed previously, infinite thickness for a given element in an analysis depends upon sample matrix and the excitation energy used. Ultimately, however, tolerance for thickness and diameter will be determined by the way in which element values are calculated. In this study, peak ratio values (i.e., Fe/Mn and Fe/Ti) show the greatest overall success. These values are statistically indistinguishable from the control group down to a thickness of 0.2 mm for Fe/Ti and to a thickness of 0.03 mm for Fe/Mn, assuming a fixed diameter of 25 mm. The difference in precision between the two ratios (see also Figs. 3.5d and 3.6d) is likely due to the low concentration of Mn relative to that of Ti in Glass Mountain obsidian (279 ppm for Mn vs. 1,601 ppm for Ti, Govindaraju, 1989). Similarly, and at a fixed thickness of 3 mm, values are indistinguishable down to a diameter of 5 mm for Fe/Mn and 3 mm for Fe/Ti, though these limits rise to 10 mm for the 1 and 2 mm thick samples.

For analysis of Ba with the X-ray tube, ppm values are indistinguishable from the control group down to a thickness of 0.2 mm. This is an excellent result given that infinite thickness for Ba in this analysis is over 1 cm. However, the same analysis seems particularly sensitive to sample diameter. Values for two of the 15 mm diameter samples (the 2 and 3 mm thicknesses) are significantly different from the control group, though the value for the 15 mm by 1 mm thick sample is not. This points to one of the limitations of this data set, which is that too few sample diameters are analyzed to fully describe the relationship between diameter and element concentration and by extension, effects of diameter and thickness combined. For now, we assume that the allowable diameter in this analysis is somewhere between 15 and 25 mm.

Analytical results for Ba acquired with the Am source are poor for all samples. Of 27 measurements, only three (the 0.2, 0.5, and 1 mm thick samples) are within 100 ppm, or 12% of the 807 ppm reported for RGM-1 (Govindaraju, 1989). Because the analyte/Compton correction usually requires samples of infinite thickness (Franzini et al., 1976), the large infinite thickness of barium at this energy may be a factor, or perhaps some other source of error is present.

Size tolerances for elements in the mid-Z analysis, as defined by comparison to the control group, are somewhat diverse. In general, the results are more conclusive, and probably more reliable, for the elements with large peaks, so this discussion will focus on Rb, Sr, and Zr. Zn and Pb are seldom used in obsidian characterization, and the poor precision of these values reflects the fact that this analysis is optimized for Rb, Sr, Y, and Zr. If required, much better results may be obtained for these elements by using a lower excitation energy. For those samples with a fixed diameter of 25 mm, element values are indistinguishable from the control group down to thicknesses of 2 mm for Rb, to 2.5 mm for Sr, and to 1.2 mm for Zr. The lower thickness limit for Zr relative to the other two elements is unexpected. As the elements are close in atomic number, they would be expected to behave similarly, relative to thickness. More likely, this difference points to inadequate sampling in the control group. In any case, substantial distortion in all three element values begins at about the 1.2 and 1 mm thicknesses (Fig. 3.5a). Results are more consistent for sample diameter. At a fixed thickness of 3 mm trace element values for all three elements are unaffected down to a diameter of 10 mm. For Sr and Zr, this threshold rises for the 2 mm thick samples, and at 1 mm thick, values for all diameters are measurably distorted.

Surface Experiment

Analytical results for the surface experiment are presented in Table 3.3. By comparing the coefficients of variation (CV) for six runs of the flaked samples to six runs of powdered samples, we expected to see a difference in precision that would reflect the error contributed by the flaked surfaces. What we see instead is that for most elements, any potential difference in precision between flaked and powdered specimens is obscured by the much larger errors contributed by peak counting uncertainty. In general, the distribution of counts in an analyte peak follows the normal distribution, and a one s error is given by the square root of counts in the peak (Jenkins et al., 1981). While other factors such as peak-to-background ratio and element sensitivity contribute to the overall precision of a measurement, the effects of counting uncertainty in this analysis are apparent in Figs. 3.7 and 3.8, where the CV for the six runs on each sample are reported by element. Element concentrations, in addition to average peak counts for each element, are given in Table 3.4.

In general, and within a given spectral region, error is larger for elements where concentration and hence, peak counts are low. This is reflected in both the magnitude and range of the CV values for elements in all samples. Coefficients of variation for Mn in the Napa Valley samples, for instance, range from 1.8 to 7.8%, where Mn concentration is an average of 167 ppm, with a peak area of approximately 2,400 counts. In the Bodie Hills samples, where Mn concentration averages 487 ppm with a peak area of approximately 8,200 counts, CV ranges from 1.5 to 2.2%.

For the flaked vs. powdered sample comparison, while lower variability was expected of powdered samples, coefficients of variation (CV) for six runs of the powdered Napa Valley and Bodie Hills samples are consistently at or below the lowest CV values for the flaked samples *only* for Fe and Zr. These elements are distinguished by large peaks and high peak-to-background ratios. Only for Fe

Table 3.3 Trace element data for 10 flaked obsidian samples (BH-1 through NV-5) and two pressed powder samples

Sample	Zn	Pb	Rb	Sr	Y	Zr	Ti	Mn	Fe	Fe/Ti	Fe/Mn	Ba	Ba(Am)
BH-1													
Mean	34	38	189	92	15	104	561	412	0.57	36	15	431	331
Min	27	35	186	90	12	101	529	403	0.56	33	15	422	323
Max	42	40	190	94	17	107	611	424	0.58	38	16	444	341
SD	5.2	2.0	1.3	1.9	1.8	2.0	24.8	6.6	0.01	1.4	0.3	8.1	5.8
CV	15.1	5.3	0.7	2.1	12.2	2.0	4.4	1.6	1.46	3.9	1.7	1.9	1.8
BH-2													
Mean	36	37	188	91	14	103	522	393	0.54	37	15	445	345
Min	30	33	183	90	11	99	493	386	0.50	35	15	433	341
Max	45	41	190	93	15	106	556	400	0.56	39	16	454	350
SD	4.9	2.8	2.8	1.1	1.4	2.7	23.6	5.1	0.01	1.4	0.3	8.6	3.1
CV	13.8	7.4	1.5	1.2	10.4	2.6	4.5	1.3	2.26	3.7	1.9	1.9	0.9
BH-3													
Mean	37	37	190	93	13	105	546	389	0.52	35	15	447	354
Min	28	36	186	91	12	101	538	377	0.51	33	15	436	345
Max	43	42	197	98	14	109	551	397	0.54	35	15	468	365
SD	4.5	2.3	3.6	2.2	1.0	2.2	4.2	8.0	0.01	0.6	0.2	10.1	7.3
CV	12.4	6.1	1.9	2.3	7.8	2.1	0.8	2.1	1.94	1.8	1.3	2.3	2.1
BH-4													
Mean	32	37	189	95	14	103	511	380	0.52	37	15	443	352
Min	29	33	184	92	12	101	487	373	0.51	36	15	436	345
Max	35	42	195	98	15	105	540	386	0.53	39	16	456	356
SD	2.2	3.1	3.6	1.9	0.9	1.3	18.7	4.9	0.01	1.2	0.1	6.5	4.7
CV	6.9	8.3	1.9	2.0	6.4	1.2	3.7	1.3	1.70	3.3	0.9	1.5	1.3
BH-5													
Mean	34	38	188	92	14	103	518	376	0.50	35	15	441	342
Min	29	36	179	91	12	97	502	364	0.49	34	15	422	338
Max	37	41	193	95	15	107	537	394	0.52	37	16	465	356
SD	2.9	1.4	5.1	1.5	1.1	3.0	11.4	10.2	0.01	0.9	0.5	13.8	6.6
CV	8.6	3.6	2.7	1.6	8.2	2.9	2.2	2.7	2.27	2.6	3.1	3.1	1.9
NV-1													
Mean	63	35	194	8	47	240	442	141	1.17	89	89	367	256
Min	58	33	192	7	45	236	416	137	1.15	82	88	354	251
Max	69	39	196	10	51	243	465	145	1.19	94	93	376	260
SD	4.2	2.0	1.3	1.0	2.1	2.3	15.2	2.4	0.02	3.8	2.0	6.7	3.4
CV	6.6	5.6	0.7	12.5	4.5	1.0	3.5	1.7	1.35	4.3	2.2	1.8	1.3
NV-2													
Mean	64	34	195	8	46	243	451	143	1.18	87	89	364	265
Min	58	32	193	6	40	239	433	129	1.16	83	83	349	258
Max	68	36	197	9	49	247	469	150	1.19	92	100	372	276
SD	3.5	1.2	1.5	1.1	2.8	2.8	11.6	7.6	0.01	2.8	5.9	8.5	6.2
CV	5.5	3.4	0.8	14.5	6.2	1.2	2.6	5.3	0.85	3.2	6.6	2.3	2.3
NV-3													
Mean	63	37	193	7	46	247	453	145	1.17	86	87	369	273
Min	56	32	190	7	43	243	433	131	1.14	82	84	362	256

(continued)

Table 3.3 (continued)

Sample	Zn	Pb	Rb	Sr	Y	Zr	Ti	Mn	Fe	Fe/Ti	Fe/Mn	Ba	Ba(Am)
Max	78	42	198	8	49	251	476	150	1.19	90	97	381	287
SD	7.6	3.2	3.0	0.5	2.3	2.7	14.3	6.6	0.02	2.8	4.5	6.2	9.8
CV	12.0	8.9	1.5	6.5	5.0	1.1	3.1	4.6	1.53	3.2	5.1	1.7	3.6
NV-4													
Mean	63	35	194	8	45	244	468	148	1.19	84	87	367	268
Min	56	32	190	7	43	241	448	128	1.17	81	80	359	264
Max	65	38	198	10	47	250	485	159	1.23	87	100	372	273
SD	3.2	2.1	2.6	0.8	1.2	3.1	12.9	11.0	0.02	2.2	7.9	5.4	2.8
CV	5.0	6.1	1.3	9.5	2.7	1.3	2.8	7.5	1.64	2.6	9.1	1.5	1.0
NV-5													
Mean	63	35	194	8	46	239	451	145	1.19	88	88	365	265
Min	56	33	191	5	45	236	416	133	1.17	83	80	348	257
Max	68	38	199	10	47	242	472	157	1.26	96	96	380	272
SD	3.4	1.8	2.8	1.4	0.7	1.9	18.6	8.1	0.03	3.9	5.2	11.0	5.2
CV	5.5	5.1	1.4	18.2	1.6	0.8	4.1	5.6	2.54	4.4	5.9	3.0	2.0
NV powder													
Mean	59	34	189	8	45	240	526	167	1.42	88	89	385	375
Min	54	30	184	6	42	237	496	155	1.41	83	86	373	332
Max	68	38	194	9	48	243	556	172	1.43	93	96	394	418
SD	4.9	2.7	3.3	1.0	2.3	2.1	20.2	6.5	0.01	3.6	3.8	6.2	39.1
CV	8.3	8.0	1.7	13.2	5.0	0.9	3.8	3.9	0.39	4.1	4.2	1.6	10.5
BH powder													
X	32	34	180	89	14	107	638	487	0.72	38	16	463	374
Min	26	31	177	87	13	105	598	479	0.72	35	15	446	342
Max	43	36	182	92	15	109	688	500	0.73	41	16	483	403
SD	5.2	1.7	1.6	1.5	0.7	1.2	27.9	6.6	0.00	1.9	0.2	11.9	27.2
CV	16.4	4.9	0.9	1.7	5.1	1.2	4.4	1.4	0.41	4.9	1.6	2.6	7.3
RGM-1 standard													
X	37	22	151	103	25	221	1665	291	2.01	36	69	803	791
Min	31	20	149	101	22	218	1638	285	2.00	35	67	794	739
Max	44	25	154	107	26	224	1687	297	2.02	37	70	819	824
SD	3.9	1.7	1.4	1.8	1.2	2.1	15.0	4.2	0.01	0.5	1.1	9.0	26.8
CV	10.5	7.7	1.0	1.7	4.8	0.9	0.9	1.4	0.39	1.3	1.7	1.1	3.4
JR-2 standard													
X	38	22	150	102	25	219	403	1000	0.80	70	8	–	–
Min	34	20	145	100	23	218	393	992	0.80	67	8	–	–
Max	42	24	152	103	28	221	415	1007	0.81	72	8	–	–
SD	3.0	1.4	2.5	1.3	1.5	1.4	7.4	4.6	0.00	1.5	0.1	–	–
CV	7.9	6.2	1.7	1.3	5.8	0.6	1.8	0.5	0.28	2.2	0.7	–	–

RGM-1 and JR-2 reference standards are also included. Statistics for all samples represent six runs in six different orientations. All element values are expressed in ppm with the exception of Fe/Mn and Fe/Ti peak ratios, and Fe ($Fe_2O_3^T$), which is expressed in weight percent. Values for Ba acquired with the [241]Am radioisotope source are labeled Ba(Am)

M.K. Davis et al.

are the CV of the powdered sample actually *lower* than the CV of the flaked samples. Furthermore, the difference between the two is quite small – a difference of 1.05 units of CV for the Bodie Hills samples and 0.46 units of CV for the Napa Valley samples (Figs. 3.7 and 3.8, Table 3.3). These data suggest that in most cases errors related to

Table 3.4 Concentration values and approximate peak counts for powdered Napa Valley and Bodie Hills samples

Element	Napa Valley		Bodie Hills	
	Concentration	Counts	Concentration	Counts
Zn	59	750	32	350
Pb	34	800	34	900
Rb	189	17,800	180	17,200
Sr	8	700	89	700
Y	45	6,600	14	1,700
Zr	240	40,600	107	16,500
Ti	526	2,800	638	3,400
Mn	167	2,400	487	8,200
Fe	1.41	238,000	1	130,000
Ba	384	32,000	463	36,000
Ba(Am)	341	4,600	407	7,200

All concentrations are in ppm, with the exception of Fe ($Fe_2O_3^T$), which is in weight percent. Values for barium acquired with the [241]Am radioisotope source are labeled Ba(Am)

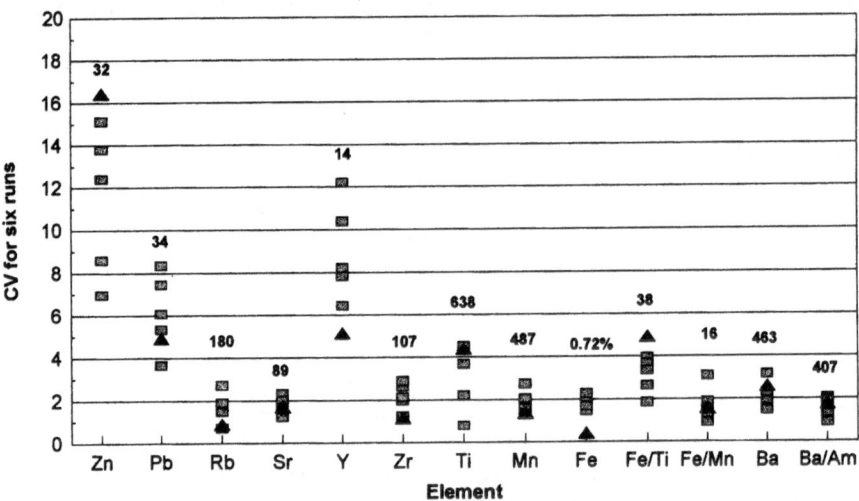

Fig. 3.7 Relative errors acquired from six runs of Bodie Hills obsidian. Each point represents the coefficient of variance (CV) of a given element in a single sample analyzed in six different orientations. Squares represent CV for unmodified samples of obsidian with flaked surfaces (BH-1 through BH-5), while the triangle represents the CV of a pressed powder sample of Bodie Hills obsidian. Numbers in bold above each element indicate the average concentration of that element. All values in ppm with the exception of Fe/Ti and Fe/Mn peak ratios, and Fe ($Fe_2O_3^T$) which is in weight percent. Values for Ba acquired with the [241]Am radioisotope source are labeled Ba/Am.

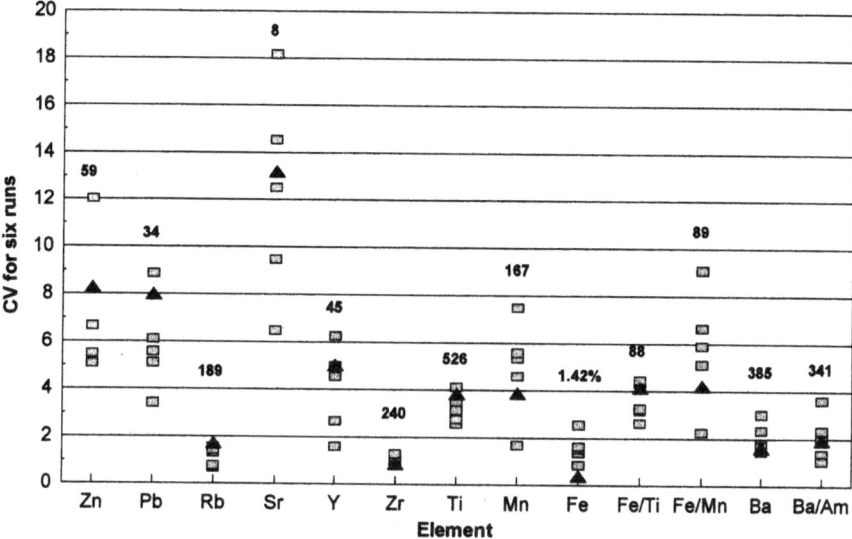

Fig. 3.8 Relative errors acquired from six runs of Napa Valley obsidian. Each point represents the coefficient of variance (CV) of a given element in a single sample analyzed in six different orientations. Squares represent CV for unmodified samples of obsidian with flaked surfaces (NV-1 through NV-5), while the triangle represents the CV of a pressed powder sample of Bodie Hills obsidian. Numbers in bold above each element indicate the average concentration of that element. All values in ppm with the exception of Fe/Ti and Fe/Mn peak ratios, and Fe (Fe$_2$O$_3$^{T}) which is in weight percent. Values for Ba acquired with the ^{241}Am radioisotope source are labeled Ba/Am

the surface topography of obsidian artifacts are negligible when compared to those contributed by counting uncertainty.

Conclusions

Two experiments were conducted to measure effects related to artifact size and surface variability in the non-destructive analysis of obsidian. On the basis of the analysis of 20 samples of Glass Mountain obsidian, the following size limits are observed. Thickness limits assume a fixed sample diameter of 25 mm, and diameter limits assume a fixed thickness of 3 mm. For Fe/Mn and Fe/Ti peak ratios element values are unaffected down to a thickness of 0.2 mm and to a diameter of 5 mm. Similarly, limits for Ba acquired with the X-ray tube are 0.2 mm for thickness and somewhere between 15 and 25 mm for diameter. Results for Ba acquired with the Am radioisotope source are poor for all samples, and we are unable to explain the source of error at this time. For the mid-Z elements, the observed thickness limit ranges between 1.2 and 2.5 mm, and the diameter limit is 10 mm.

Archaeologists requiring the analysis of small artifacts will want to note that the size limits reported here reflect the points to which no statistically measurable element distortions are observed. Such a level of precision is rarely required for accurate source ascription of obsidian artifacts, and element values for many of the smaller samples in this set reflect only minor error. Furthermore, relative element proportions often remain intact, even for significantly undersized samples (e.g., 0.5 mm thick by 8 mm in diameter), and may be accurately characterized in some cases depending upon the context of the analysis. It is up to the archaeologist and the analyst to determine the analytical precision required to distinguish sources or chemical groups in a given study area.

Results of the surface experiment, where samples of flaked Napa Valley and Bodie Hills obsidian were analyzed in an attempt to measure error related to surface irregularities, suggest that this error is nearly always negligible when compared to counting uncertainty. The only instance where precision for powdered samples (CV over six runs of one sample) differs visibly from those of flaked samples is for the element Fe, where the peak is large and peak-to-background ratio is high (Figs. 3.7 and 3.8, Table 3.3). In this case, precision for the powdered samples improved over those of the flaked samples by an average of less than 1% CV for both the Bodie Hills and Napa Valley samples.

Regarding the applicability of these results to the analysis of artifacts, all samples in this study were prepared from freshly broken and clean obsidian, which is, needless to say, uncommon in the world of obsidian artifacts. Effects due to soil contamination and weathering are not considered here. Secondly, the flaked samples used in the surface experiment are relatively flat so that the convexity often encountered in flaked stone tools is not represented. Finally, these results are specific to the analytical conditions outlined herein and may or may not be comparable to results from other laboratories. For this reason, it is important that analytical methods and conditions be reported completely with the XRF results so that comparability may be evaluated.

References

Bouey, P., (1991), Recognizing the limits of archaeological applications of non-destructive energy-dispersive X-ray fluorescence analysis of obsidians. *Materials Research Society Proceedings* 185, 309–320.

Davis, M.K., (1994), Bremsstrahlung ratio technique applied to the non-destructive energy-dispersive X-ray fluorescence analysis of obsidian. *International Association for Obsidian Studies Bulletin* 11.

Franzini, M., Leoni, L., and Saitta M., (1976), Determination of the X-ray fluorescence mass absorption coefficient by measurement of the intensity of Ag Kα Compton scattered radiation. *X-ray Spectrometry* 5:84–87.

Govindaraju, K., (1989), 1989 Compilation of working values and sample description for 272 geostandards. *Geostandards Newsletter* 13 (special issue).

Jenkins, R., Gould, R.W., and Gedcke, D., (1981), *Quantitative X-ray Spectrometry*. New York: Marcel Dekker

Hughes, R.E., (1984), Obsidian source studies in the Great Basin: Problems and Prospects. In Hughes, R.E., Ed., *Obsidian Studies in the Great Basin*, (pp. 1–20). Berkeley: Contributions of the University of California Archaeological Research Facility 45.

Hughes RE (1988), The Coso Volcanic field reexamined: implications for obsidian sourcing and hydration dating research. Geoarchaeology 3:253–265

Jackson, T.L. and Hampel, J.H., (1992), Size effects in the energy-dispersive X-ray fluorescence (EDXRF) analysis of archaeological obsidian artifacts. Presented at the 28th International Symposium on Archaeometry, Los Angeles.

McCarthy, J.J. and Schamber, F.H., (1981), Least-squares fit with digital filter: a status report. In Heinrich, K.F.J., Newbury, D.E., Myklebust, R.L. and Fiori, E. Eds., *Energy Dispersive X-ray Spectrometry*, (pp. 273–296). Washington, D.C.: National Bureau of Standards Special Publication 604.

Schamber, F.H., (1977), A modification of the linear least-squares fitting method which provides continuum suppression. In Dzubay, T.G., Ed., *X-ray Fluorescence Analysis of Environmental Samples*, (pp. 241–257). Ann Arbor: Ann Arbor Science.

Shackley MS (1988), Sources of archaeological obsidian in the Southwest: an archaeological, petrological, and geochemical study. American Antiquity 53:752–772

Shackley,M.S. and Hampel, J., (1992), Surface effects in the energy dispersive X-ray fluorescence (EDXRF) analysis of archaeological obsidian. Presented at the 28th International Symposium on Archaeometry, Los Angeles.

Tatlock, D.B., Flanagan, F.J., Bastron, H., Berman, S., and Sutton, A.L. Jr., (1976), Rhyolite, RGM-1, from Glass Mountain, California. In Flanagan, F.J., Ed., *Descriptions and Analyses of Eight New USGS Rock Standards*. U.S. Geological Survey Professional Paper 850, (pp. 11–14).

Chapter 4
Non-destructive EDXRF Analyses of Archaeological Basalts

Steven P. Lundblad, Peter R. Mills, Arian Drake-Raue and Scott Kekuewa Kikiloi

Introduction

Lundblad et al. (2008) addressed methodological considerations for nondestructive energy dispersive X-ray fluorescence (EDXRF) of archaeological basalts by examining the effects of sample size and thickness on geochemistry, and by preliminarily addressing the effects of weathering on archaeological basalt. Those findings reinforce and expand on the promising EDXRF results obtained by others for basalts (Jackson et al. 1994; Latham et al. 1996; Northwest Research Obsidian Studies Laboratory 2008; Weisler & Kirch 1996; Weisler 1993a, b). Basalt samples larger than 1 cm in diameter and more than 1 mm thick generally produce reliable source characterization geochemistry and mirror the findings obtained by Davis et al. (1998) (Chap. 3 here) for obsidians. Here, we expand on those central points by presenting additional summary data defining three methodological issues affecting results: (1) chemical weathering on archaeological basalts, (2) surface contamination by phosphates, and (3) surface morphology and textural variation. Data is mainly presented as summaries from datasets we have collected in Hawai'i in order to highlight the effects these factors have on measured geochemistry.

In the Pacific Islands, geochemical characterization of archaeological basalts for provenance studies has been practiced for over two decades (Bayman and Moniz Nakamura 2001; Best 1984; Collerson & Weisler 2007; Weisler 1998, 1997, 1990, 1993a, b; Weisler and Woodhead 1995; Winterhoff 2003). Collerson and Weisler (2007) argued that isotopic ratios may be the most reliable way to determine the island source of basalt artifacts, but because this technique is destructive, time-consuming, and relatively expensive, there are many compelling reasons to continue using EDXRF as a first tier archaeometric approach to large sample groups (Mills et al. 2008). EDXRF continues to offer non-destructive and broad-scale sampling that

S.P. Lundblad (✉)
Department of Geology, University of Hawaii-Hilo, Hilo, HI 96720, USA
e-mail: slundbla@hawaii.edu

M.S. Shackley (ed.), *X-Ray Fluorescence Spectrometry (XRF) in Geoarchaeology*,
DOI 10.1007/978-1-4419-6886-9_4, © Springer Science+Business Media, LLC 2011

would not be feasible with ICP-MS, isotope ratio, or other expensive and destructive techniques.

EDXRF analysis of basalts, however, involves several challenges that are less problematic with obsidians. Major element and trace element concentrations in basalts tend to be more heterogeneous than in obsidian and also exhibit less geographic distinctiveness because of the more continuous and expansive nature of mafic eruptions. Major Polynesian basalt quarry sites have been characterized and compared (Sinton and Sinoto 1997; Mills et al. 2008), but minor sources with similar geochemical signatures, such as cobbles from gulches or dense basalt from dikes, confound our ability to make exclusive associations with specific sources. In the Hawaiian Islands, hot spot volcanism is responsible for widespread distribution of geochemically related flows from the same magma source, and elemental concentrations between flows exhibit strong covariance. Geochemical trends are generally repeated as volcanoes evolve from tholeiitic shield-building phases through postshield alkalic eruptions. There have been a number of extensive geochemical datasets published for Hawai'i (Cousens et al. 2003; Casadevall and Dzurisin 1987a, b; Frey et al. 1990; Garcia et al. 1992, 2000, 2003; Moore et al. 1987; Rhodes 1996; Tilling et al. 1987; Wolfe et al. 1997; to name a few), but these studies are not focused on the specific flows that Hawaiians used to make tools. In addition to continued characterization of source material, analyses of local geology in the vicinity of cultural sites is essential to adequately characterize and discriminate local and nonlocal sources.

EDXRF has the methodological advantage of providing large-scale and non-destructive analysis of many elements that can be used to characterize basalt artifacts at the scale of eruptive phases. There are several challenges, however, when using these elements to ascribe an artifact to a specific source. Non-destructive EDXRF analysis of basalt requires a thorough understanding of the factors affecting measured geochemical concentrations at the time of analysis. This study focuses on factors affecting the surface of samples, necessarily a component of any non-destructive analysis.

Basalt weathers rapidly in highly acidic tropical soils subjected to abundant rainfall on windward sides of the islands. This is largely a leaching process where acids selectively dissolve different compounds in the rock. It is thus expected that trace and major element concentrations and ratios will be altered by chemical weathering. It is consequently important to determine the potential extent of chemical alteration that can be expected for different basalts, especially when conducting nondestructive analyses of artifacts recovered in acidic soils.

Surface patination or contamination is a similar issue to acidic leaching of basalts, but it is additive rather than reductive. One potentially significant surface contaminant in archaeological midden and in coastal sites is phosphate. Guano from birds and bats, as well as phosphates from archaeological midden can patinate basalts, and the overall effect of this process on surface geochemistry needs to be established.

Finally, textural variation in basalt artifacts is common. While most adzes were made from fine grained and dense basalts, other classes of artifacts were made from vesicular basalts such as poi pounders, abraders, sinkers, and stone images. The texture of a single Hawaiian lava flow can vary from highly porous vesicular basalts to dense 'a'a flows. In non-destructive EDXRF analyses, these textural changes will affect the peak intensities for the measured elements, and it is consequently important to determine the effects of this variable on analytical precision.

The Hilo Method

The University of Hawai'i at Hilo's ThermoScientific QuanX™ EDXRF spectrometer is primarily devoted to the archaeometric study of Oceanic basalts and volcanic glasses. Full details of our analytical method can be found in Lundblad et al. (2008). Currently, Hilo's EDXRF spectrometer is the only one in Hawai'i that is committed to cultural research. A customized large sample chamber accommodates artifacts, such as adze blanks, poi pounders, and stone bowls.

The QuanX uses a Rhodium (Rh) stable-isotope X-ray tube, thermoelectrically-cooled detector, and supporting Edmunds vacuum pump, with data processed on Wintrace™ software, version 3.1, build 33. Our analytical technique focuses on a suite of 17 elements ranging in atomic weight from magnesium (Mg) to niobium (Nb). Analysis of the lightest of these elements requires a vacuum environment, although heavier elements can be analyzed in normal air environments. We complete analyses for all elements under vacuum conditions. Elemental concentrations for the heavier elements are calculated using ratios of the peak intensities for the heavier elements and the Compton scatter background. Background scatter is lower in the vacuum condition, which significantly improves resolution.

We affix 4 μm thick Ultralene™ X-ray film over the sample wheel apertures on which lithic samples are placed. This allows us to analyze samples as small as 10 mm in diameter, the size below which measured geochemical concentrations significantly change (Lundblad et al. 2008). It also allows us to optimally position smaller samples over the X-ray beam. To maintain a standardized technique, we run large samples and small samples with the same film in place.

Calibration

In order for EDXRF spectrometers to conduct quantitative analyses, the spectrometers need to be calibrated by analyzing similar geological reference standards with well-established concentrations of elements. The UH Hilo spectrometer has been calibrated for the analysis of basalt with 27 geological standards (see Lundblad et al. 2008 for details). All these standards are distributed in powdered form. We reconstitute the powders into pellets using a 2% polyvinyl alcohol binder, pressing them in a 25-ton hydraulic press to better correspond with our whole-rock analyses. The pellets are made in a 31 mm dye to match the apertures on the QuanX sample wheels.

An important consideration when conducting spectrometer calibrations is the placement of the pellets in the same position as the samples that one intends to analyze. The QuanX sample wheels are manufactured so that 31 mm pellets can be seated *within* the sample wheel apertures. However, archaeological samples are generally placed *above* the apertures, 4 mm higher than the pellet position. Change in height above the X-ray source affects beam intensity and therefore alters the concentration data. This offset distance results in measurable differences for our calibration. Only if the calibration and analysis height are the same will the measured and accepted values be equivalent.

Factors Influencing Measured Geochemistry

Chemical Weathering of Archaeological Basalt

Lundblad et al. (2008) compared weathered surface geochemistry with the interior geochemistry for a single adze blank recovered from Pololu Adze Quarry (50-10-03-4981) on the northern tip of Hawai'i Island from which a thin-section had been made, exposing a cut surface on the sample (Tuggle 1976; Lass 1994). Debitage at this site was buried in thick deposits next to a stream with little soil development. Although there was some discoloration of the artifact's surface from weathering, no significant difference in geochemistry for target elements was observed while comparing the weathered surface and the freshly exposed surface. This initial finding demonstrated that even artifacts recovered in windward valleys subjected to centuries of weathering can provide reliable geochemical characterizations.

To further test the effects of weathering on archaeological specimens, we analyzed basalt flakes recovered from the plow-zone of a former sugarcane field in the windward community of O'okala on Hawai'i Island. The area is underlain by an O'okala silty loam, with relatively highly acidity (pH = 5.0), and is subjected to an annual rainfall of over 100 in. per year (USDA 2008). The extent of weathering on the flakes from O'okala is more extreme than that observed on the Pololu sample as determined by surface color and weathering rind thickness. Weathering on O'okala artifacts appears pale gray to grayish-brown (Munsell Soil Color Chart 2.5Y 5/0 to 2.5Y 5/2) and contrasts with dark gray interiors (2.5Y 5/4) on freshly exposed surfaces.

Twenty-one flakes that were larger than 2 cm in diameter and had relatively flat ventral surfaces were analyzed. The ventral surfaces were then ground down to expose the dark black interiors of the flakes, and the samples were reanalyzed. Individual elements show systematic change between weathered and fresh surfaces. For the Mid-Z elements, Rb, Zr, and Nb increased in concentration as a result of weathering, while Sr and Y decreased. Overall, the effect of weathering changes the value for the centroid calculated using PCA for all elements as shown in Fig. 4.1 and Table 4.1. The measured change, however, is small when compared to the overall variation measured for source areas in Hawai'i.

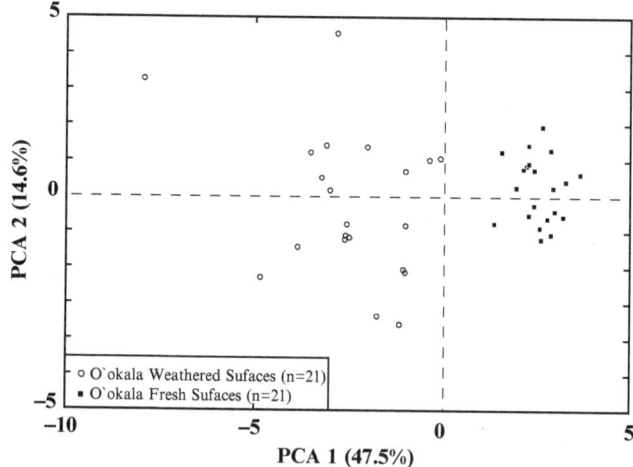

Fig. 4.1 Bivariate plot of sample scores on first and second principal components illustrating differences in weathered and fresh surfaces from O'okala, Hawaii. PC1 represents 47.5% of the observed variation between the groups, and PC2 represents 14.6% of the observed variation

Table 4.1 Eigenvectors for the first five principal components describing weathering effects at O'okala

Element	PC1	PC2	PC3	PC4	PC5
MgO (%)	0.270	0.008	−0.375	−0.119	0.264
Al$_2$O$_3$ (%)	−0.288	0.195	0.007	−0.116	−0.204
SiO$_2$ (%)	0.078	0.333	−0.158	−0.788	0.009
K$_2$O (%)	−0.318	−0.123	0.193	−0.074	−0.115
CaO (%)	0.271	0.091	0.014	−0.103	−0.586
TiO$_2$ (%)	−0.323	−0.081	0.142	−0.035	−0.162
V (ppm)	−0.319	0.022	0.137	−0.057	−0.183
MnO (ppm)	0.163	−0.014	0.465	0.019	−0.224
Fe (%)	−0.293	0.164	−0.158	0.074	0.079
Ni (ppm)	0.207	−0.365	0.218	−0.152	0.071
Cu (ppm)	0.250	0.143	0.129	0.333	0.132
Zn (ppm)	−0.148	−0.070	−0.539	0.312	−0.288
Rb (ppm)	−0.272	−0.059	0.012	−0.064	−0.115
Sr (ppm)	−0.023	0.590	−0.030	0.113	−0.087
Y (ppm)	0.086	0.454	0.376	0.128	0.115
Zr (ppm)	−0.279	0.213	0.055	0.109	0.411
Nb (ppm)	−0.240	−0.180	0.130	−0.220	0.328

Chemical Weathering of Geological Basalt

To better understand the effect of weathering on the measured compositions of Hawaiian basalts, we analyzed 60 bedrock geological samples from one of the known basalt quarries of the Haleakala volcano on Maui (Site 50-50-11-2510) which was extensively used between AD 1400 and 1600 (Carson and Mintmier 2006). These samples were cut and analyzed on both the cut, flat side, and on the weathered surface. The degree of weathering on these slabs is generally greater and more variable than that on archaeological basalt due to longer exposure and more variable weathering conditions. Recognizing that intense chemical weathering over long time periods in Hawai'i would be heavily selected against in archaeological basalt, it is nevertheless important to assess how weathering changes the measured geochemistry of basalts in the region. Weathering was qualitatively assessed in these samples using a three-part scale. Samples fell into the following groups: Group I showed minor weathering with little color change from the original; Group II had significant oxidation with noticeable change from the original color; Group III is characterized by more extreme weathering than Group II due to the presence of well-developed weathering rinds.

The 60 geological samples from this site comprise a single geochemical cluster, distinct from material analyzed from Hawai'i Island and other Maui sources thus far at UH-Hilo. When the geochemical data from weathered and fresh surfaces are compared, three observations are clear. One, the data from weathered surfaces has significantly greater scatter, as indicated by increased standard deviation, than that from the fresh surfaces. Two, there is a significant change in mean value for some of the measured elements. Three, much of the variation is attributable to the samples with the highest degree of weathering. Samples exhibiting Group III weathering account for most of the change in geochemistry. Figure 4.2 shows this relationship for Sr and Zr. Strontium is one of the only elements for which this high degree of weathering changes the measured concentration in a consistent manner. For the remaining elements, the highly weathered samples do not show consistent change in their measured geochemistry.

Removing the most weathered samples (Group III) from the dataset reveals that there is no significant difference in compositional means between populations for TiO_2, MnO, Ni, Rb, Sr, Y, Zr, or Nb ($p < 0.01$) between Groups I and II (Fig. 4.3a). Major element geochemistry means differ significantly for Al_2O_3, SiO_2, K_2O, and CaO (Fig. 4.3b). This is likely a function of the lower X-ray beam intensity used for these elements, which sample a higher proportion of the weathered outer surface relative to the unaltered interior. In general, weathering does not affect the measured average composition of Mid-Z trace elements except in extreme cases, whereas the scatter in the data increases in all cases (Table 4.2). This type of increased uncertainty in the data could lead to confusion between two closely related geochemical populations. Fortunately, there is enough geochemical variation between the major known quarry sites in Hawai'i that even extreme weathering does not render cluster analysis useless (Sinton and Sinoto 1997).

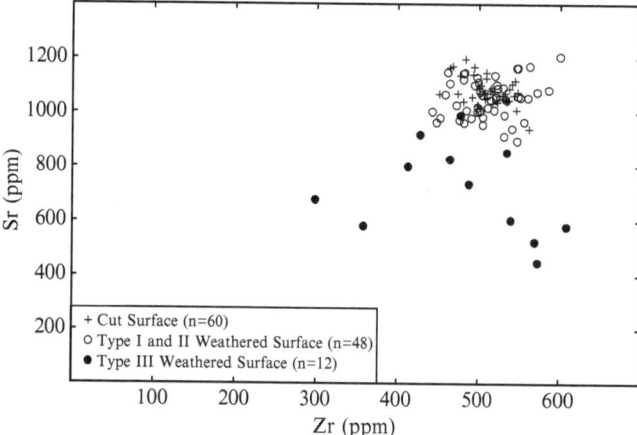

Fig. 4.2 Bivariate plot of strontium (Sr) and zirconium (Zr) for weathered and cut geologic samples from Haleakala Volcano, Maui. Fresh surfaces and those displaying minor weathering show no significant variation, while intensely weathered surfaces are significantly reduced in measured concentration for Sr

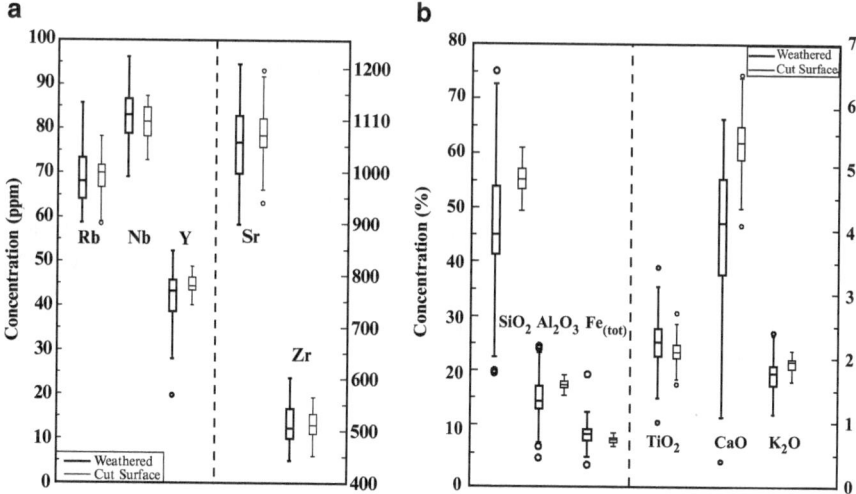

Fig. 4.3 Box plots of trace element (**a**) and major element (**b**) data for geologic samples from Haleakala Volcano. *Boxes* represent 50% of the data between the 25 and 75% quartile. Maximum and minimum values lie at the end of the bars, and suspected outliers are represented by *circles*. Median values for each element are shown with the *line* within the boxes. Note scale changes on each diagram

Table 4.2 Comparison of weathered and cut geologic samples from Site 50-50-11-2510. Weathered samples have been subdivided as described in the text

	Weathered I II ($n = 48$)	SD	Weathered III ($n = 12$)	SD	Cut surface ($n = 60$)	SD
Al$_2$O$_3$ (%)	14.9	4.2	14.0	7.5	17.4	0.9
SiO$_2$ (%)	49.5	16.3	59.1	29.3	55.0	2.9
K$_2$O (%)	1.7	0.3	1.2	0.4	1.9	0.1
CaO (%)	4.0	1.1	2.4	1.2	5.4	0.4
TiO$_2$ (%)	2.3	0.4	1.8	0.5	2.1	0.2
V (ppm)	206	33	179	53	169	28
MnO (ppm)	3,078	555	6,925	4,916	3,171	214
Fe (%)	8.4	2.5	7.8	5.1	7.2	0.6
Ni (ppm)	19	17	30	36	20	6
Cu (ppm)	14	7	40	25	10	7
Zn (ppm)	150	18	188	76	140	6
Rb (ppm)	68	8	73	35	69	4
Sr (ppm)	1,052	74	712	169	1,075	49
Y (ppm)	42	5	39	7	45	2
Zr (ppm)	512	37	480	93	512	25
Nb (ppm)	83	6	76	15	81	4

Surface Contamination by Phosphates: Examples from the Northwest Hawaiian Islands

While weathering is one form of alteration that can affect archaeological basalt, salts from sea spray and phosphates from fecal material can pervade and/or bond with basalts, thereby affecting the measured geochemistry of a sample. An interesting example of this phenomenon is found in the Northwest Hawaiian Islands. These isolated islands extend northwestward from the main Hawaiian Islands as a series of deeply eroded volcanic islands and coral atolls. The island of Nihoa, located approximately 250 km northwest of Kaua'i, was settled sometime between AD 1000 and AD 1700, with a number of known archaeological sites and associated artifacts (Emory 1928; Kirch 1985). The island is also home to a large number of birds which results in abundant phosphate deposition on the island. The bird concentration is due, at least in part, to the lack of other landmasses in the area. While generally not a problem in destructive geochemical analysis, surface contaminants pose a potential difficulty for nondestructive EDXRF. In order to assess the impact of phosphate, and to a lesser extent salt (NaCl), on the measured geochemistry, we analyzed artifacts directly from the field without any attempt to clean them, and compared that information with samples ultrasonically cleaned in a dilute HCl solution. Despite the absence of visible contamination on the artifacts, we detected noticeable P and Cl peaks during analysis for the light (Mid-Za condition as defined in the Wintrace software) elements (Fig. 4.4a). This was

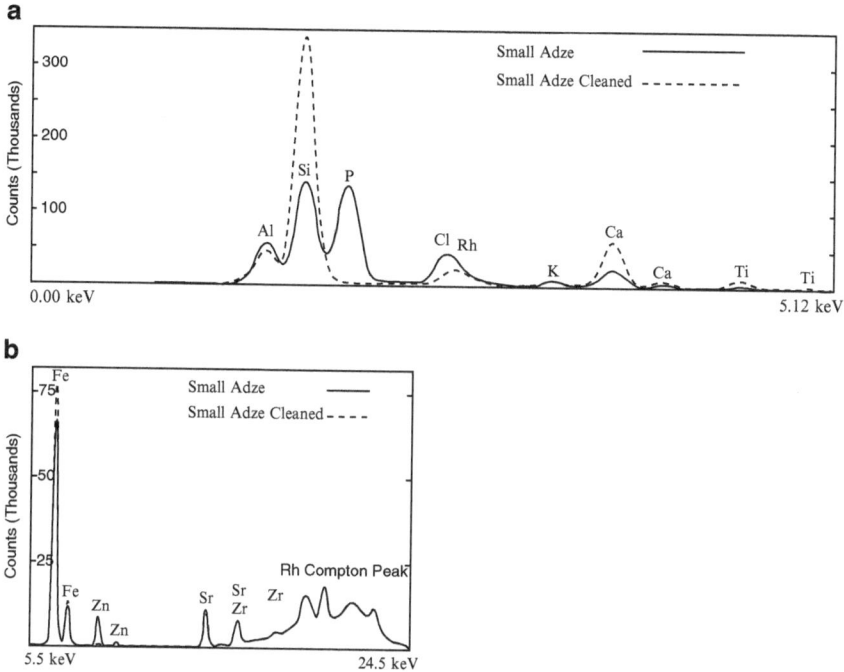

Fig. 4.4 Comparison of X-ray spectra from a small Nihoa Island Adze showing variation in measured X-ray intensity due to the presence of phosphate contaminants (*solid line*) when compared to a clean sample (*dashed line*). (**a**) Low-Za condition (no filter, 6 kV, vacuum). (**b**) Mid-Zc (0.125 mm Pd filter, 28 kV, vacuum)

accompanied by a corresponding reduction in the number of counts and measured concentration in most of the major elements, most notably SiO_2, which measured less than 50% of the counts from a clean sample (Table 4.3).

When the same artifact was analyzed at higher excitation energy for the Mid-Zc trace elements (Rb, Sr, Y, Zr, Nb), no measurable difference was identified in concentration and the spectra for the two cases were very similar (Fig. 4.4b). The only notable spectral exception for this analytical condition was Ni.

Surface Morphology and Textural Variation

To minimize the effects of surface irregularities, some non-destructive analyses have relied on ratios of various trace elements to identify geological groups (Latham et al. 1992; Weisler and Kirch 1996). Ratios, however, can obfuscate real quantitative differences in sources if, in fact, elemental concentrations covary. The amount of error introduced from minor surface irregularities, such as a flaked basalt adze surface, has been insignificant (Lundblad et al. 2008). Consequently, we

Table 4.3 Comparison of original and cleaned Nihoa adze (50-NH-60) showing the effect of phosphate contamination on measured counts per second (cps) and calculated geochemistry

Element	Adze concentration	cps	Cleaned Adze concentration	cps
Al_2O_3 (%)	2.4	192	11.9	739
SiO_2 (%)	2.9	0	59.4	503
K_2O (%)	0.7	49	0.8	45
CaO (%)	1.9	253	7.6	244
TiO_2 (%)	1.4	239	4.0	240
V (ppm)	182	16	445	17
MnO (ppm)	446	38	1,150	38
Fe (%)	6.9	7,364	8.2	7,290
Ni (ppm)	74	18	154	18
Cu (ppm)	107	12	68	13
Zn (ppm)	1,598	421	133	500
Rb (ppm)	16	25	16	22
Sr (ppm)	552	983	485	964
Y (ppm)	31	60	28	63
Zr (ppm)	228	490	238	490
Nb (ppm)	19	46	20	43

have attempted to conduct initial characterization of sources with elemental concentrations rather than ratios.

Debitage and Changes in Surface Morphology

We tested for the effect of sample shape and corresponding sample height changes on measured geochemical composition and analytical variability by analyzing a large number of cut geological samples and archaeological flakes from the Mauna Kea Adze Quarry Complex (Mills et al. 2008). Geologic samples were collected from the main bedrock locations within the site, cut into slabs, and each surface analyzed. Additionally, we analyzed 820 flakes from four rock shelters within the quarry complex ('Mills and Lundblad, 2006; Mills et al. 2008). Each flake was run in a position to optimize its excitation by the X-ray source, and the flattest noncortical surface chosen. While the flakes were collected from four rockshelters that represented different parts of the quarry (mainly Bishop Museum Sites 2 and 14, with additional flakes from Sites 7 and 11), they do not constitute a representative sample of the quarry complex. Consequently, as expected, the mean values for the flakes and the geologic slabs differ. When the variation in the data is compared, however, an interesting trend appears. Standard deviations for the Mauna Kea Adze Quarry Complex flake and slab datasets are not significantly different for any of the Mid-Z trace elements except for Sr. Eight flakes produced outlier geochemistry, perhaps due to extremely irregular surfaces. Even with these outliers removed from the dataset, there is still a significant increase in the amount of variation for Sr. This indicates that Sr is more susceptible to alteration caused by surface irregularities. Major element components are more susceptible to surface variability with CaO,

Table 4.4 Summary of measured geochemistry for geologic samples analyzed on a flat surface and archaeological debitage (flakes) from the Mauna Kea Adze Quarry Complex

	Cut slab average (n = 495)	SD	Flake average (n = 820)	SD
MgO (%)	3.9	0.7	2.2	1.1
Al$_2$O$_3$ (%)	13.4	0.5	12.5	1.5
SiO$_2$ (%)	49.7	1.4	45.7	3.7
K$_2$O (%)	0.8	0.1	0.9	0.1
CaO (%)	8.4	0.4	8.5	0.5
TiO$_2$ (%)	3.5	0.2	3.6	0.2
V (ppm)	448	27	438	22
MnO (ppm)	1,649	63	1,768	296
Fe (%)	10.7	1.0	11.1	0.7
Ni (ppm)	32	6	29	9
Cu (ppm)	67	23	55	24
Zn (ppm)	141	6	150	17
Rb (ppm)	26	4	30	4
Sr (ppm)	558	13	568	25
Y (ppm)	41	3	43	3
Zr (ppm)	306	22	320	20
Nb (ppm)	32	3	35	3

MnO, SiO$_2$, and MgO demonstrating increased variation, while Al$_2$O$_3$, Fe, and K$_2$O did not vary significantly (Table 4.4).

Textural Variation and Vesicular Basalt

While much of our research has been focused on the dense basalt typically used in making adzes, other stone artifacts were constructed from more porous basalt. Dense basalt appropriate for making tools such as adzes is not found everywhere in Hawai'i, and these sources provide geochemical fingerprints that are distinctive and can be identified as markers for artifacts even after transportation throughout the archipelago. Vesicular basalt, on the other hand, is much more common and consequently is more difficult to ascribe to an exact source location. In order to assess the relationship between vesicular and dense basalt analyses, we compared the measured geochemistry of 10 vesicular midden rock samples collected from Kahalu'u Habitation Cave (Site 50-10-37-7702) with pressed pellets made from these samples after they were initially analyzed. The pressed pellets better approximate the density and surface texture of naturally-occurring dense basalt. Vesicular samples showed greater scatter in measured concentration as measured by standard deviation of the samples compared to the same samples analyzed as pressed pellets. Major elements showed significant variation between the two groups. As expected, vesicular samples consistently produced fewer counts, presumably due to higher scattering of the X-ray beam, and corresponding lower concentrations for MgO, Al$_2$O$_3$, SiO$_2$, TiO$_2$, and MnO,

Table 4.5 Comparison of vesicular rock samples and pressed pellets from the Kahalu'u Habitation Cave

	Rock average	SD	Pellet average	SD
MgO (%)	5.2	2.7	13.6	2.1
Al$_2$O$_3$ (%)	9.1	2.0	11.5	1.7
SiO$_2$ (%)	34.5	9.7	44.1	2.9
K$_2$O (%)	0.8	0.5	0.6	0.1
CaO (%)	11.0	1.2	10.1	0.4
TiO$_2$ (%)	1.6	0.2	1.7	0.1
V (ppm)	229	35	253	20
MnO (ppm)	1,515	121	1,681	53
Fe (%)	12.0	2.6	10.4	0.7
Ni (ppm)	99	19	216	21
Cu (ppm)	198	118	63	19
Zn (ppm)	167	42	81	22
Rb (ppm)	19	9	16	6
Sr (ppm)	298	21	279	15
Y (ppm)	18	2	17	2
Zr (ppm)	83	13	79	6
Nb (ppm)	15	2	15	2

while K$_2$O, CaO, and Fe measured concentrations were higher for vesicular samples (Table 4.5).

Concentrations of Mid-Z trace elements were not significantly different, indicating that they remain reasonably good indicators of overall geochemistry. The variability in the measured composition for these elements, however, increases due to the surface irregularities present in vesicular rocks. Trace element geochemistry can therefore be used to help determine the relationship between vesicular rocks and their dense basalt counterparts. Without an accurate geochemical assessment of the major elements, assigning vesicular samples to specific sources or associating them with dense basalt artifacts remains somewhat uncertain.

Conclusions

EDXRF can be an extremely useful analysis tool for examining large collections of lithic material that might otherwise be off-limits due to cost, time, or cultural constraints. However, data can be properly interpreted only if we understand the factors affecting non-destructive analysis. In contrast to most destructive techniques, size, surface morphology, and weathering can influence the measured geochemical composition of samples when using non-destructive EDXRF.

The role of weathering is a significant concern for any nondestructive analysis in which a weathered surface is included. Based on analysis of weathered rocks from O'okala (Hawai'i Island) and Haleakala (Maui Island), the variability in measured geochemical composition increases for analyses conducted on weathered surfaces.

Major elements, analyzed at lower X-ray intensities, show more variation than the Mid-Z trace elements (Rb, Sr, Y, Zr, Nb), making them better indicators of alteration, but poorer indicators of original chemical composition. Silica, for example, is a very sensitive element to changes in surface irregularities, weathering, and contamination by phosphate. Because SiO_2 concentration in virtually all Hawaiian lavas range between 40 and 63%, and the majority of lavas have compositions between 45 and 50% erupted as part of their shield building stage (Clague and Dalrymple 1987), lower SiO_2 values are not an accurate measure of original chemical composition, but can function as a "canary in a coal mine." The Mid-Z elements, on the other hand, are relatively unaffected by weathering except in extreme cases. This, coupled with their utility as good discriminators of sources in Hawai'i, make them excellent targets for analysis.

Preliminary analysis of surface texture influence (vesicles) and surface morphology (cuspate flakes), indicates a similar pattern to that found with weathering, with the measured compositions varying significantly for many of the major elements yet remaining relatively constant for the Mid-Z trace elements. Analytical variation is significantly greater for all elements.

The greater uncertainty in measured geochemistry due to weathering, surface morphology, or weathering can be overcome by analyzing a greater number of samples. While clusters of data points may be more diffuse for samples that have higher degrees of weathering, their mean values are not significantly altered, and consequently the overall geochemistry can be determined.

Acknowledgments Acquisition of the EDXRF at the University of Hawai'i-Hilo was supported by a major research instrumentation grant from the National Science Foundation (BCS 0317528). We gratefully acknowledge Randy Cone (ThermoNoran), Craig Skinner (Northwest Research Obsidian Studies Laboratory), and Steven Shackley (UC-Berkeley) for their assistance with EDXRF methods. We also thank Melanie Mintmier and Haleakala National Park for obtaining samples.

References

Bayman, J. M., & Moniz Nakamura, J. J. (2001). Craft specialization and adze production on Hawai'i Island. *Journal of Field Archaeology, 28*, 239–252.

Best, S. B. (1984). *Lakeba: The prehistory of a Fijian island*. Unpublished doctoral dissertation, University of Aukland.

Carson, M. T., & Mintmier, M. A. (2006). *Archaeological site documentation in front country areas in the summit district of Haleakala National Park, Maui Island, Hawaii*. Honolulu: International Archaeological Research Institute.

Casadevall, T. J., & Dzurisin, D. (1987). Stratigraphy and Petrology of the Uwekahuna Bluff Section, Kilauea Caldera. *USGS Professional Paper, 1350*, 351–376.

Casadevall, T. J., & Dzurisin, D. (1987). Intrusive Rocks of Kilauea Caldera. *USGS Professional Paper, 1350*, 377–394.

Clague, D. A., & Dalrymple, G. B. (1987). The Hawaiian-Emperor Volcanic Chain Part I: Geologic Evolution. *USGS Professional Paper, 1350,* 5–54.

Collerson, K. D., & Weisler, M. I. (2007). Stone adze compositions and the extent of ancient Polynesian voyaging and trade. *Science, 317,* 1907–1911.

Cousens, B. L., Clague, D. A., & Sharp, W. D. (2003). Chronology, chemistry, and origin of trachytes from Hualalai Volcano, Hawaii. *Geochemistry Geophysics Geosystems, 4(9),* 27. doi: 10.1029/2003GC000560.

Davis, M. K., Jackson, T. L., Shackley, M. S., Teague, T., & Hampel, J. H. (1998). Factors affecting the energy dispersive X-ray fluorescence (EDXRF) analysis of archaeological obsidian. In M. S. Shackley (Ed.), *Archaeological obsidian studies: method and theory.* (pp. 59–80). New York: Plenum.

Emory, K. P. (1928). *Archaeology of Nihoa and Necker Islands, Tanager expedition no. 5. Bishop Museum Bulletin,* 53.

Frey, F. A., Wise, W. S., Garcia, M. O., West, H. B., Kwon, S., & Kennedy, A. K. (1990). Evolution of Mauna Kea Volcano, Hawaii; petrologic and geochemical constraints on postshield volcanism. *Journal of Geophysical Research, 95(B2),* 1271–1300.

Garcia, M. O., Rhodes, J. M., Ho, R. A., Ulrich, G. E., & Wolfe, E. W. (1992). Petrology of lavas from episodes 2-47 of the Pu'u 'O'o eruption of Kialuea Volcano, Hawaii: evaluation of magmatic processes. *Bulletin of Volcanology, 55,* 1–16.

Garcia, M. O., Pietruszka, A. J., Rhodes, J. M., & Swanson, K. (2000). Magmatic processes during the prolonged Pu'u 'O'o eruption of Kilauea Volcano, Hawaii. *Journal of Petrology, 41,* 967–990.

Garcia, M. O., Pietruszka, A. J., & Rhodes, J. M. (2003). A petrologic perspective of Kilauea Volcano's summit magma reservoir. *Journal of Petrology, 44,* 2313–2339. doi: 10.1093/petrology/egg079.

Jackson, R. J., Jackson, T. J., Miksicek, C., Roper, K., & Simons, D. (1994). *Framework for archaeological research and management: National Forests of the North-Central Sierra Nevada, Unit III: special studies and research data.* Santa Cruz: BioSystems Analysis.

Kirch, P. V. (1985). *Feather gods and fishhooks.* Honolulu: University of Hawaii Press.

Lass, B. (1994). *Hawaiian adze production and distribution: implications for the development of chiefdoms.* (UCLA Institute of Archaeology Monograph No. 37). Los Angeles: University of California Los Angeles.

Latham, T., Sutton, P. A., & Versub, K. L. (1992). Non-destructive XRF characterization of basaltic artifacts from Truckee, California. *Geoarchaeology, 7,* 81–101.

Lundblad, S. P., Mills, P. R., & Hon, K. (2008). Analysing archaeological basalt using nondestructive energy-dispersive X-ray fluorescence (EDXRF): effects of post-depositional chemical weathering and sample size on analytical precision. *Archaeometry, 50,* 1–11.

Mills, P. R., & Lundblad, S.P. (2006). *Preliminary field report: the geochemistry of the Ko'oko'olau Complex, Mauna Kea Adze Quarry (50-10-23-4136) TMK: 4-4-15:10.* Hilo: University of Hawaii at Hilo, Geoarchaeology Laboratory.

Mills, P. R., Lundblad, S. P., Smith, J. G., McCoy, P. C., & Nalemaile, S. P. (2008). Science and sensitivity: a geochemical characterization of the Mauna Kea Adze Quarry Complex, Hawaii Island, Hawaii. *American Antiquity, 73,* 743–758.

Moore, R. B., Clague, D. A., Rubin, M., & Bohrson, W. A. (1987). Hualalai Volcano: A Preliminary Summary of Geologic, Petrologic, and Geophysical Data. *USGS Professional Paper, 1350,* 571–586.

Northwest Research Obsidian Studies Laboratory. (2008). *Non-Obsidian Geochemical Characterization References.* Retrieved Sept 21, 2008, from: http://www.obsidianlab.com/basalt/basalt_universe.html.

Rhodes, J. M. (1996). Geochemical stratigraphy of lava flows sampled by the Hawaii Scientific Drilling Project. *Journal of Geophysical Research, 101(B5),* 11729–11746.

Sinton, J. M. & Sinoto, Y. H. (1997). A geochemical database for Polynesian adze studies. In M. I. Weisler (Ed.), *Prehistoric long-distance interaction in Oceania: an interdisciplinary approach.* (pp. 194–204). Auckland: New Zealand Archaeological Association Monograph 21.

Tilling, R. I., Wright, T. L., & Millard Jr., H. T. (1987). Track-element chemistry of Kilauea and Mauna Loa lava in space and time: a reconnaissance. *USGS Professional Paper 1350,* 641–689.

Tuggle H. D. (1976). *Windward Kohala-Hamakua archaeological zone, island of Hawaii.* Honolulu: University of Hawaii, Manoa, Department of Anthropology.

United States Department of Agriculture. (2008). *Soil and Climate Data.* Retrieved Sept 21, 2008, from: http://ortho.ftw.nrcs.usda.gov/osd/dat/O/OOKALA.html.

Weisler, M. I. (1990). Sources and sourcing of volcanic glass in Hawaii: implications for exchange studies. *Archaeology in Oceania, 25,* 16–25.

Weisler, M. I. (1993). Provenance studies of Polynesian Basalt Adze Material: a review and suggestions for improving regional data bases. *Asian Perspectives, 32,* 61–83.

Weisler, M. I. (1993b). Chemical characterization and Provenance of Manu'a Adz Material using a non-destructive x-ray fluorescence technique. In P. V. Kirch & T. L. Hunt (Eds.), *The To'aga Site: Three millenia of Polynesian occupation in the Manu'a Islands, American Samoa.* (pp. 167–187). Berkeley: Contributions of the University of California Archaeological Research Facility 51.

Weisler, M. I. (Ed.). (1997). Prehistoric long-distance interaction in Oceania: an interdisciplinary approach. *New Zealand Archaeological Association Monograph, 21.*

Weisler, M. I. (1998). Hard evidence for prehistoric interaction in Polynesia. *Current Anthropology 39,* 521–532.

Weisler, M. I., & Kirch, P. V. (1996). Interisland and interarchipelago transfer of stone tools in prehistoric Polynesia. *Proceedings of the National Academy of Sciences, 93,* 1381–1385.

Weisler, M. I., & Woodhead, J. D. (1995). Basalt Pb isotope analysis and the prehistoric settlement of Polynesia. *Proceedings of the National Academy of Sciences, 92,* 1881–1885.

Winterhoff, E. Q. (2003). *Ma'a mai Malaeloa: A geochemical investigation of a newly discovered basalt quarry source on Tutuila, American Samoa.* Unpublished master's thesis, University of Oregon.

Wolfe, E. W., Wise, W. S., & Dalrymple, G. B. (1997). *The geology and petrology of Mauna Kea Volcano, Hawaii – a study of Postshield volcanism* (U.S. Geological Survey Professional Paper 1557). Washington: U.S. Government Printing Office.

Chapter 5
Non-destructive Applications of Wavelength XRF in Obsidian Studies

Annamaria De Francesco, M. Bocci, and G.M. Crisci

Introduction

During the Neolithic period, the Mediterranean area represented a very important and ancient exchange community for prehistoric populations. In this period, the major activity was certainly the lithic industry, and surely the production and the use of obsidian artifacts promoted the circulation of both raw materials and artifacts in the entire Mediterranean area. Provenance studies have undergone a strong development in the last decades, also thanks to the increase in collaboration between researchers of historical and scientific traditions.

The possible geological sources of obsidians in the Mediterranean area are located on the islands of Lipari, Pantelleria, Sardinia, Palmarola, and the Greek islands of Melos and Gyali. Different destructive or non-destructive analytical methods have been proposed to identify the area of geological origin of obsidians (Cann and Renfrew 1964; Francaviglia 1984; Hallam et al. 1976; Thorpe et al. 1984; Birò et al. 1986; Thorpe 1995; Bigazzi et al. 1993; Acquafredda et al. 1996, 1999; Kayani and McDonnel 1996; Tykot and Young 1996, 1997; Shackley 1998a, b, 2002, 2005; Gratuze 1999; Stewart et al. 2003; Bellot-Gurlet et al. 2004; Glascock and Neff 2003; Summerhayes et al. 1998; Cirrincione et al. 1995). All the methods used to determine the composition of natural glasses, such as obsidians, are based on methods requiring the physical and/or chemical transformation of the samples and/or costly analytical procedures such as OES (Optical Emission Spectrography), EDX-RF (Energy Dispersion X-ray Fluorescence), WDXRF (Wave Dispersion X-ray Fluorescence), SEM-EDS, Neutron Activation, Fission Tracks, Electron Microprobe, Inductively Coupled Plasma Mass Spectrometer (ICP-MS) associated with a Laser Ablation, Mossbauer Spectrometer, PIXE/PIGME (proton-induced X-ray emission/proton-induced gamma ray emission spectrometry), and portable XRF. Neutron

A. De Francesco (✉)
Dipartimento di Scienze della Terra, Università della Calabria, 87036 Rende (CS), Italy
e-mail: defrancesco@unical.it

M.S. Shackley (ed.), *X-Ray Fluorescence Spectrometry (XRF) in Geoarchaeology*,
DOI 10.1007/978-1-4419-6886-9_5, © Springer Science+Business Media, LLC 2011

activation analysis (NAA), for example, yields very good analytical results of the raw sample. However, it presents other drawbacks like the isolation of the probes due to the radioactivity induced by the activating processes and the extremely high cost of each analysis (see Glascock, Chap. 8 here).

As Mediterranean obsidians are easily distinguishable from a compositional point of view, the principal obstacle has been methodological. In fact, in this type of research, it is very important to have a non-destructive inexpensive analytical method of easy application, which allows the analysis of all finds recovered in excavations, without any limitation in choice; this can, however, sometimes exclude particularly meaningful finds. To address these issues, a non-destructive analytical methodology using WDXRF has been proposed by Crisci et al. (1994) and optimized by De Francesco et al. (2000, 2005, 2008a).

In the present work, this methodology has been further verified for all the obsidian of the Mediterranean area, by characterization of the chemical composition and of the geochemical variability of the different obsidian sources, which were initially established through a traditional XRF analysis on powders. However, the principal aim of this work is the comparison between the X-ray fluorescence methodology on powder (destructive) and the non-destructive XRF methodology on fragments (splinters of obsidian) for archaeometrical purposes.

This methodology, which has been widely used for numerous Italian Neolithic archaeological sites, allows one to go back to the origin of the archaeological obsidians of the whole Mediterranean area (De Francesco et al. 2002a, b, 2004, 2005, 2006; De Francesco and Crisci 2003; Antonelli et al. 2002, 2006; Bietti et al. 2004; Campetti et al. 2001; Langella et al. 2003).

The present work represents, therefore, not only a verification of the sensibility of the nondestructive XRF methodology in distinguishing between the different obsidian sources, but also a further control of the precision of the nondestructive XRF methodology in the attribution of the origin of archaeological obsidians.

Geological Sources of Obsidian in the Mediterranean Area

The obsidian sources in the Mediterranean area are located on the Italian islands of Sardinia, Lipari, Palmarola, Pantelleria, and the Greek islands of Melos and Gyali (Fig. 5.1).

The outcrops of obsidian on such islands are very numerous, but not all of them own technical properties for the production of artifacts. For example, on the island of Antiparos in the Greek archipelago, obsidian of scarce quality is present and has, therefore, never been used before as raw material.

The region of Monte Arci in Sardinia has been characterized by different volcanic episodes dating back to the Plio-Pleistocene. The first episode generated a large amount of sub-alkaline rhyolites, without maphic mineral inclusion, but also sub-alkaline rhyolites with maphic inclusion (Montanini et al. 1994). The obsidians have been dated with different geochronological methods: $^{40}Ar/^{39}Ar$ values indicate 3.16–3.24 MY (Montanini and Villa 1993).

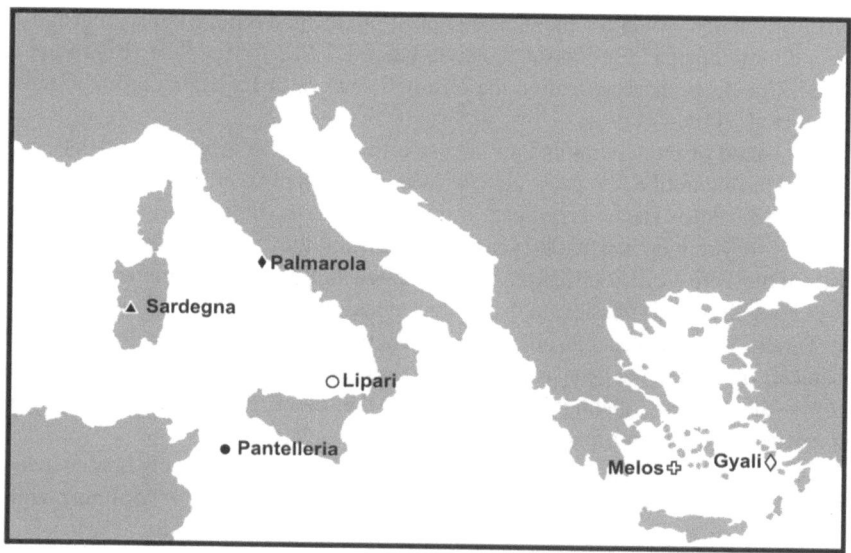

Fig. 5.1 The distribution in the Mediterranean area of obsidian source that was exploitable in the Neolithic period

In studies on the provenance of archaeological obsidians in Sardinia, Hallam et al. (1976) and Machey and Warren (1983) found three possible sources of obsidian in the volcanic complex of M. Arci, which were characterized by different compositions: Conca Cannas (SA), Santa Maria Zuarbara (SB), and Perdas Urias (SC). Using different methodologies, Thorpe et al. (1984), Tykot (1997), De Francesco and Crisci (1999), and De Francesco et al. (2004) distinguished four to five geochemical groups in the obsidians of M. Arci (SA, SB1, SB2, SC1, and SC2). This last distinction (between SC1 and SC2) is poorly significant for archaeological purposes. Tykot (2002) recognized seven chemically distinguishable groups on the basis of very slight differences in the chemical element compositions.

Lipari, which is the major Eolian island, had magmatic activity 230,000 years ago that lasted until the Roman epoch (580), characterized by a succession of ten cycles (Crisci et al. 1991) each of which with release of initial pyroclastic products and with final castings of lava flow.

On the island of Lipari, there are more obsidian castings of good quality, but the only one from the Neolithic period that has been exploited is that of the Vallone Gabellotto-Fiume Bianco, which is fission-track dated to maximum 11,400 and 8,600 years BP (Bigazzi and Bonadonna 1973) and is visible as blocks in a thick deposit of pyroclasts.

Palmarola is the westernmost of the Pontine Islands Archipelago in the Gulf of Gaeta. It is formed by calc-alkaline volcanic rocks of Pliocene age. The formation of sodic rhyolitic domes happened almost 1.7 MY ago (Barberi et al. 1967). The obsidian of Palmarola is both in primary and secondary deposits, namely as castings or as blocks. It outcrops along the southern slope of M. Tramontana and next to Punta Vardella (as secondary deposits). Macroscopically, two varieties are distinguishable, which are nearly similar for chemical composition;,the first is more transparent and the second more opaque. Fission track data indicate 1.7 MY for the flow (Bigazzi et al. 1971). Chemical characterization the obsidian sources of Palmarola are reported in Tykot et al. 2005.

The island of Pantelleria had an intense effusive activity beginning 50,000 years ago and lasting until 8,000 years ago, divided in six principal eruptive cycles (Civetta et al. 1984, 1988). The emerged part of the island and the largest part of the obsidian deposits are younger than 50,000 years, which means, they were formed soon after the issue of the Tufo Verde Formation, which is dated 49,600 years (Cornette et al. 1983).

In Pantelleria, it is possible to distinguish at least three geochemical groups: Balata dei Turchi, Salto La Vecchia, and Lago di Venere (Francaviglia 1988). Its peralkaline character (related to the genesis through fractioned crystallization of basaltic magma) allows them to be distinguished from all the other Mediterranean obsidians; Zr permits an efficient separation for the different sources.

Concerning the Greek islands, the obsidians of Gyali are extremely homogeneous and well distinguishable (Cann and Renfrew 1964; Renfrew et al. 1965; Francaviglia 1984), both from those Greek and from all the Mediterranean ones. The obsidian domes and rhyolitic lava flows are located in the north-eastern parts of the island (Di Paola 1974). The fission track method yielded an age of 30 Ka (Bigazzi and Radi 1981). Melos is the other island of the Aegean Archipelago with obsidian of good quality. Rhyolitic domes and lavas with obsidians are related to Lower Pleistocene activity (Fyticas et al. 1976, 1986). The obsidian is located mainly in two principal outcrops, Aghia Nikia and Demenegakion, compositionally homogeneous but distinguishable for the slightly different Zr content (Francaviglia 1984).

XRF Powder Methodology

The first stage of the work consisted of the characterization through analysis on powder of 65 samples of geological obsidians, which were directly sampled from the principal obsidian sources of the Mediterranean area.

At the Dipartimento di Scienze della Terra dell'Università della Calabria, all samples were chemically analyzed through WDXRF using a Philips PW 1480 Spectrometer, and the routine procedure on powders.

Major and trace elements (Nb, Sr, Zr, Rb, Y, Ni, Cr, Ce, La, Co, and V) were determined, using the procedure for the correction of matrix effects of Franzini et al. (1972, 1975) and of Leoni and Saitta (1976).

Non-destructive XRF Methodology on Entire Fragments

The non-destructive XRF analytical methodology has been optimized at the Dipartimento di Scienze della Terra dell'Università della Calabria (Crisci et al. 1994; De Francesco et al. 2000)

Fig. 5.2 The sample container of the PW 1480 XRF and an integral obsidian fragment positioned for non-destructive analysis

Sixty-five obsidian samples were sampled from the geological outcrops (already analyzed on powders with the XRF routine analysis), with morphology similar to workshops usually found in the archaeological sites (Fig. 5.2). The best surface (more regular and flat) is blocked with a sticker on the sample support, placed in the sample container, and analyzed thereafter.

The non-destructive XRF analytical methodology consists of irradiating the entire sample of obsidian with primary X-rays; as a result, the chemical elements present in the sample are excited and produce secondary X-rays (see Chap. 2). The type and the intensity of secondary rays are dependent on the type and concentration of the excited chemical elements. As with all XRF analyses, for a specific chemical element, the intensity of the emitted X-rays depends, besides its concentration, also on other factors, such as:

(a) Absorption effects due to the presence of major elements in the sample
(b) Granulometrical effects due to the dimension and form of crystals within whole rock samples
(c) Particularly for WXRF, surface effects of the X-ray irradiated and analyzed samples

In WXRF sample treatment, it is necessary to create a perfectly flat irradiation surface and homogenize the sample. The homogenization is necessary as a rock is often constituted of minerals with variable distribution and granulometry. This big theoretical limitation is absent in the case of most obsidian, which is a volcanic glass and is, therefore, perfectly homogeneous. In addition, the possible effects of absorption are negligible. The only theoretical obstacle to the use of the WXRF analysis for archaeological obsidians is only of geometric nature and is related to the fact that the irradiation surface is not flat and varies from sample to sample. This is a particular issue when the peak heights are not ratioed to the Compton or bremsstrahlung scatter. For this reason, only surface effects avoid the direct transformation of secondary X-ray intensities in absolute concentrations. If we consider that the radiation emitted from two similar and contemporarily analyzed chemical elements of the same sample, create similar variations, the intensity ratios of such elements can be used as real concentrations and, therefore, used to build discriminating plots. The dimension of the analyzed fragments must be between 1 and 5 cm because of the XRF irradiation container. However, the irradiated area may be focused relative to the fragment size (circular area of 1 or 2 cm of diameter). As noted above, only five chemical elements, Nb, Y, Zr, Rb, and Sr, have been selected as they are more than enough to characterize the different provenance areas.

Results of the Chemical Analyses on Powders

The mean values of anhydrous chemical analyses on powders, of the 65 samples of Mediterranean obsidians, are listed in Table 5.1. All major elements and many trace elements (Nb, Sr, Zr, Rb, Y, Ni, Cr, Ce, La, Co and V) were analyzed. For the classification of the obsidians, the Total Alkali/Silica (TAS) of the Le Bas et al. (1986) system was used (Fig. 5.3a), indicating that all samples were rhyolite.

Fig. 5.3b clearly shows the groupings corresponding to the different sources. The results, in the silica–alkali plot are partially overlapped only for Sardinia, Lipari, and Palmarola. Melos and Gyali are well separated due to the high value of SiO_2 and a lower content of alkali. The obsidians of Pantelleria are separated notably from all the others due to their peralkaline character, as mentioned above.

The diagram in Fig. 5.4a (Nb ppm vs. Zr ppm) shows how the obsidians of Monte Arci (Sardinia) are classified in four groups, SA, SC, SB1, and SB2, which are well established in the literature. The obsidians sampled near Conca Cannas and Uras are clustered in the SA group, those sampled near Pau, Perdas Urias, and Sennixeddu in the SC group, and those from Santa Maria Zuarbara and Marrubiu in the SB1 and SB2 group. Figure 5.4a shows that, based on the values of Zr and Nb, the obsidians of Palmarola, Lipari, Melos, and Gyali are in well-separated fields, but from a geochemical point of view, they are nevertheless homogeneous. The obsidians of Melos show slightly different Zr contents among the two outcrops of the island (Aghia Nikia and Demenegakion). The high Zr content of the obsidians of Pantelleria (the only ones from the Mediterranean that are formed through fractional crystallization from basaltic melts) determines the clear isolation of such source from the rest of the Mediterranean sources; besides, it is possible to distinguish at least three geochemical groups of obsidians: Lago di Venere, Salto La Vecchia, and Balata dei Turchi (at higher values of Zr).

The same groups are also evident in the diagram of Fig. 5.4b (Zr ppm vs. Sr ppm), where Sr indicates the differences already observed in previous diagram, between the obsidian sources of the Mediterranean area. Palmarola and Pantelleria show the lowest values of Sr content, but in the diagram they occupy distant areas due to the Zr content. Sardinia's four groups are easily distinguished due to the varied Sr content, similar to Lipari and the Greek islands.

The referenced diagrams represent two examples that indicate how a few trace elements (only Zr, Sr and Nb used here) are sufficient to distinguish between the obsidian sources of the entire Mediterranean area with the application of

Table 5.1 Anhydrous chemical analysis of all major elements and some trace elements of obsidian samples from the Mediterranean sources. Major elements are expressed in weight %; trace elements in ppm (parts per million)

	SA (15)		SC (11)		SB1 (6)		SB2 (3)		Gyali (4)		Palmarola (8)		Lipari (4)		Pant. SL (2)		Pant. BT (2)		Pant. LV (2)		Milos (8)	
	Media	Dev. st.	Media	Dev. st.	Media	Dev. st.	Media	Dev. st.	Media	Dev. st.	Media	Dev. st.	Media	Dev. st.	Media	Dev. st.	Media	Dev. st.	Media	Dev. st.	Media	Dev. st.
SiO_2	75.06	0.07	73.05	0.22	74.33	0.55	75.49	0.11	78.41	0.71	74.47	0.17	75.62	0.60	74.99	2.49	71.22	0.96	69.20	0.64	77.09	0.21
Al_2O_3	13.56	0.03	14.01	0.09	13.59	0.23	13.12	0.03	11.68	0.38	13.10	0.06	12.54	0.32	6.64	0.72	7.60	0.03	10.41	0.26	12.67	0.13
Fe_2O_3	1.58	0.03	2.01	0.06	1.84	0.15	1.53	0.01	1.03	0.04	1.90	0.07	1.84	0.07	7.75	0.66	9.18	0.83	7.71	0.65	1.18	0.01
MnO	0.06	0.00	0.04	0.01	0.05	0.01	0.04	0.00	0.04	0.01	0.08	0.01	0.07	0.01	0.26	0.02	0.31	0.02	0.29	0.01	0.06	0.00
MgO	0.26	0.01	0.46	0.02	0.36	0.04	0.28	0.00	0.23	0.01	0.19	0.02	0.18	0.01	0.11	0.00	0.14	0.02	0.33	0.04	0.36	0.01
CaO	0.66	0.01	0.98	0.02	0.82	0.09	0.63	0.01	0.63	0.02	0.52	0.01	0.74	0.03	0.25	0.03	0.32	0.04	0.48	0.07	1.35	0.03
Na_2O	3.50	0.02	3.29	0.03	3.46	0.04	3.38	0.01	3.66	0.10	4.77	0.10	4.01	0.07	6.14	0.60	6.85	0.01	6.51	0.35	3.79	0.04
K_2O	5.15	0.02	5.73	0.04	5.25	0.05	5.33	0.06	4.18	0.15	4.84	0.11	4.91	0.13	3.56	0.33	4.09	0.11	4.50	0.18	3.31	0.03
TiO_2	0.11	0.00	0.33	0.01	0.20	0.03	0.15	0.00	0.13	0.01	0.11	0.01	0.09	0.01	0.29	0.13	0.28	0.03	0.55	0.03	0.17	0.01
P_2O_5	0.06	0.00	0.11	0.01	0.08	0.01	0.06	0.00	0.01	0.00	0.01	0.00	0.01	0.00	0.02	0.00	0.02	0.00	0.04	0.00	0.03	0.00
Nb	56	1	30	1	45	7	30	1	17	1	72	3	39	4	280	33	339	16	223	7	8	1
Y	37	1	24	1	30	5	21	1	13	1	74	2	44	4	167	14	199	5	127	4	15	2
Rb	257	2	175	2	245	2	246	1	137	5	399	16	288	7	156	15	177	2	126	8	117	3
Zr	96	1	213	3	132	17	120	9	103	3	263	8	147	21	1467	140	1720	21	1207	138	111	12
Sr	31	1	134	3	68	13	40	4	69	1	9	1	19	2	6	0	11	4	6	0	105	3
Ni	3	1	3	1	4	1	3	0	3	0	3	1	4	2	3	0	3	1	15	18	3	3
Cr	3	1	3	1	3	1	3	1	4	1	2	1	5	3	6	2	5	0	9	6	4	2
V	3	1	14	2	8	3	2	0	3	1	1	1	2	1	1	1	2	2	5	1	6	1
La	26	4	70	4	35	3	31	3	43	3	98	4	67	9	230	18	256	5	163	23	28	3
Ce	57	4	132	7	73	3	65	0	64	8	185	6	126	18	413	25	468	20	299	33	47	6
Co	3	1	4	1	3	1	3	0	2	1	3	1	3	0	12	0	13	1	11	2	3	1
Ba	127	4	899	19	255	39	164	38	780	31	31	7	29	3	69	4	80	18	74	23	474	7

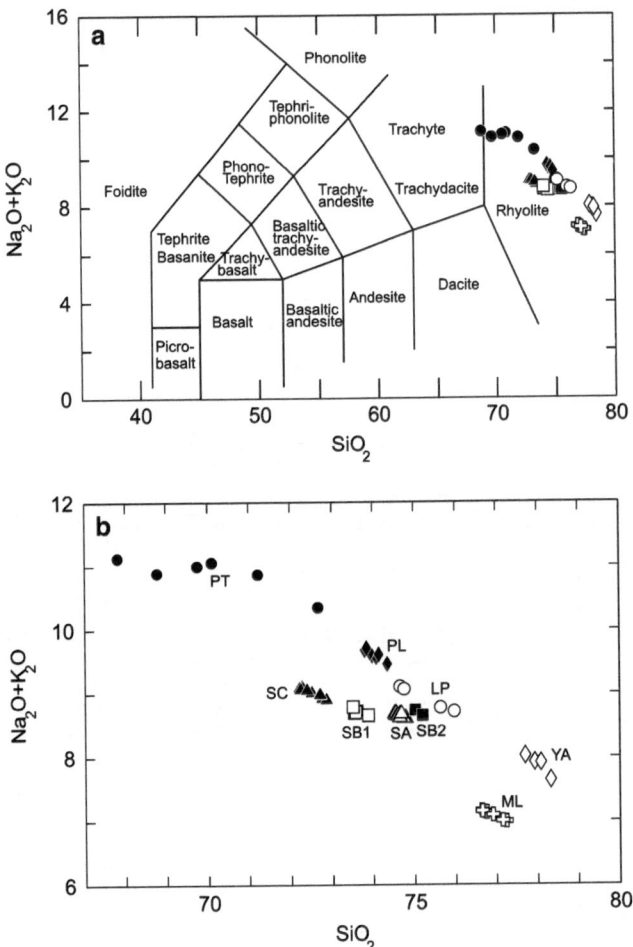

Fig. 5.3 (a) The classifying diagram of Total Alkali-Silica (Le Bas et al., 1986). All the obsidians belong to the rhyolitic rocks field. (b) A detail of the same diagram. Close to the symbols, the legend is shown in the diagram: *PT* Pantelleria; *PL* Palmarola; *SA*, *SB1*, *SB2* and *SC* Sardegna – M. Arci; *LP* Lipari; *ML* Melos; *YA* Gyali

non-destructive XRF methodology. Other trace elements, such as Rb, Y, Ba, Ce, and La, are also very useful for the same purpose.

Comparison with the Analyses on the Entire Fragments

From a methodological point of view, we present a comparison between the trace (ppm) elements acquired from Mediterranean obsidian in powders and the second-ary X-ray intensities of the same chemical elements analyzed in entire fragments of

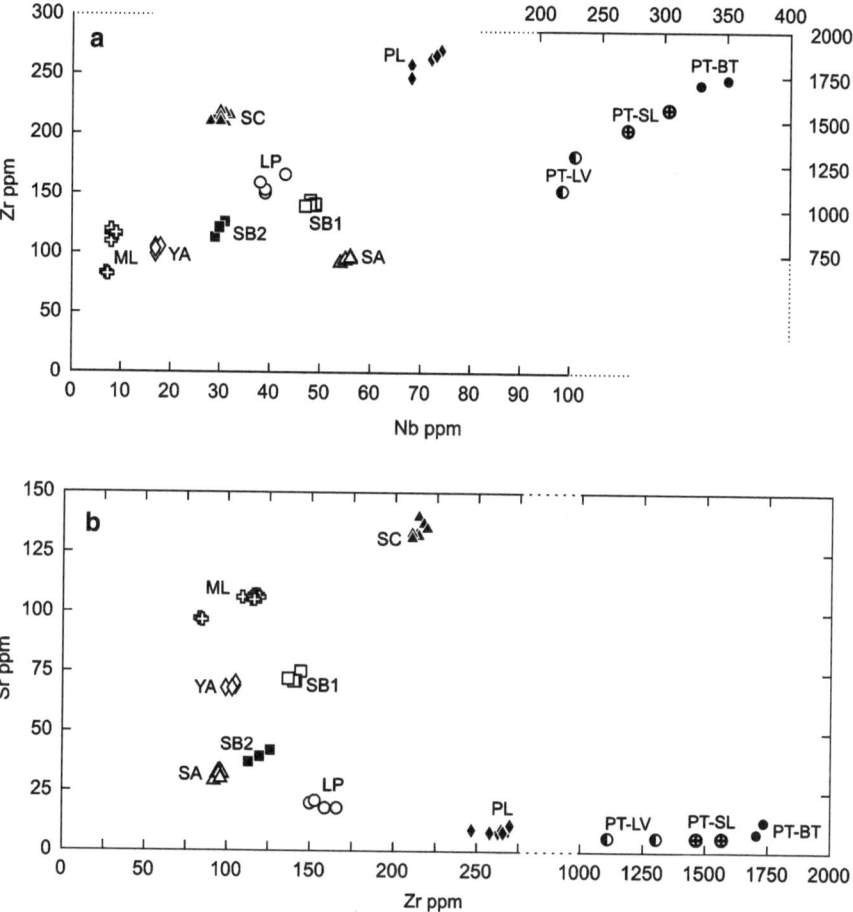

Fig. 5.4 (a) The Nb vs. Zr binary diagram, in ppm. Legend as in Fig. 5.3. (b) The Zr vs. Sr binary diagram, in ppm. Legend as in Fig. 5.3

the same samples. The secondary X-ray intensities are reported in Table 5.2. As mentioned earlier, because of the impossibility of correcting surface effects (linked to the shape and irregularity of the radiating surface of the entire fragments, which are also, in contrast to the perfectly plane surface, different for each fragment, in the routine analyses on powders), the X-ray intensity ratios of only five chemical elements (Nb, Y, Zr, Rb, and Sr), have been used; they are more than enough to characterize the different provenance areas.

Based on this, it follows that the comparison between the two methodologies must be demonstrated between the ratios of the concentrations in ppm and the X-ray intensity ratios of the same chemical elements. Some plots, representative of the comparison between the two methodologies, are presented in Fig. 5.5.

Table 5.2 Secondary X-rays intensity of the five selected trace elements obtained on the geological splinters using non-destructive methodology

Sample	Nb	Y	Rb	Zr	Sr
SA-01	18250	6058	53002	22806	5185
SA-02	16803	5662	50706	21248	4834
SA-03	17930	5989	52064	22125	5181
SA-04	14726	4431	44326	19177	4039
SA-05	14946	4650	44850	18075	4245
SA-06	14172	4277	42570	18027	4047
SA-07	18084	5941	54042	24005	5333
SA-08	14721	4478	45128	18525	4255
SA-09	16814	5397	50208	20673	4643
SA-10	13496	3730	40783	16258	3703
SC-01	7734	3426	28871	46190	16863
SC-02	8292	3668	30549	47909	18339
SC-03	8739	3611	32318	53398	19906
SC-04	7620	3402	28735	47242	17048
SC-05	7561	3164	28369	46710	18668
SC-06	9062	3881	33307	54042	20036
SC-07	6565	2899	23769	39207	14285
SC-08	9089	3915	34353	52776	20470
SC-09	10077	4322	34929	53912	20649
SC-10	9737	4189	34340	53824	20919
SB1-01	10722	3635	36128	25891	8405
SB1-02	15132	4944	50256	32030	12917
SB1-03	13356	4322	45406	29348	10758
SB1-04	12188	3922	37829	25235	11705
SB1-05	12795	4922	42155	28245	9911
SB1-06	11214	3781	38198	25825	10395
SB1-07	14983	4764	46276	28365	10599
SB1-08	10813	3946	36127	25250	8822
SB1-09	12281	4450	42305	27303	9264
SB1-10	13582	4619	44794	28924	10267
SB2-01	7372	2929	38973	21978	4389
SB2-02	7853	2784	40224	26036	7447
SB2-03	7881	3063	42107	22453	4656
SB2-04	8982	3445	46490	27643	6405
SB2-05	9105	3239	47177	28272	7856
SB2-06	10320	4138	53047	28422	5612
SB2-07	7037	2519	36074	21155	5273
SB2-08	7756	3036	40877	22259	4608
SB2-09	8625	3550	45968	29168	8200
YA01	4352	3980	19172	17345	8220
YA02	3517	3004	13720	13903	5535
YA03	3517	2942	13923	13951	5714
YA04	3150	2998	14815	12945	5324
YA05	4277	3602	18535	16742	6596
YA06	3852	3247	15784	15136	6431
YA07	4215	3726	19113	16841	7439

(continued)

Table 5.2 (continued)

Sample	Nb	Y	Rb	Zr	Sr
YA08	2971	2465	11890	11337	4967
YA09	3110	2652	12495	12311	5067
YA10	3347	2664	13014	13056	5213
PALM-01b	11272	9226	46214	38068	678
PALM-01c	14596	12271	59065	46582	770
PALM-01d	14174	11907	59619	48426	962
PALM-01e	12416	10430	50808	41093	590
PALM-01g	14706	12505	60814	48845	636
PALM-01h	14638	12203	59560	47741	815
PALM-01m	16163	13113	65035	52257	744
PALM-01n	16043	12792	64386	51346	834
PALM-01o	14564	11664	59039	47300	768
LIP-LC01	6840	7563	34712	26620	1697
LIP-LC02	8370	8827	40396	30955	2246
LIP-LCD01	8319	9367	40888	31806	2022
LIP-LCD02	8337	8943	40721	31755	1887
LIP-RIN01	8180	9383	40980	32539	1997
LIP-RIN02	7558	8596	37689	29964	2028
LIP-VL02	6331	6604	30746	25450	1606
LIP-VL04	5540	5691	26550	21634	1321
LIP-VL05	6880	7460	34477	27832	1734
LIP-VL06	6082	6908	31585	25719	1504
PANT-03	42767	9872	12461	149773	333
PANT-04	41729	9361	11801	143908	383
PANT-07	55071	12704	15612	189818	380
PANT-BT01	62603	13376	17069	267110	492
PANT-BT02	50962	11657	14244	175297	390
PANT-BT03	57260	13174	16116	262745	421
PANT-LV01	33095	7143	9786	109187	179
PANT-LV02	34758	7572	10138	113666	195
PANT-LV03	56937	13256	16313	197950	409
MILOS-01	2191	3447	15855	20623	11701
MILOS-03	2132	3345	15926	20212	11525
MILOS-04	2366	3424	16173	20834	11976
MILOS-05	2234	3369	15883	20118	11643
MILOS-06	2068	2931	14275	17853	10484
MILOS-07	2353	3526	16101	20237	11932
MILOS-08	2161	3331	15265	19407	11171
MILOS-09	2219	3413	15555	19946	11500
MILOS-10	2342	3170	15295	19644	11035

Figure 5.5a, Rb/Sr vs. Zr/Y, refers to the ratios in ppm of the chemical elements (methodology on powders) and shows how, using the concentration ratios, the groups of obsidians of the Mediterranean area, Lipari, Palmarola, Pantelleria, Sardinia – Monte Arci (SA, SC, SB1, and SB2), and the Greek islands of Melos and Gyali are well separated. The diagram Rb/Sr vs. Zr/Y in Fig. 5.5b is shown with the intensity ratios (non-destructive methodology on entire fragment). This diagram clearly indicates the presence of similar groups, perfectly distinct and comparable with those

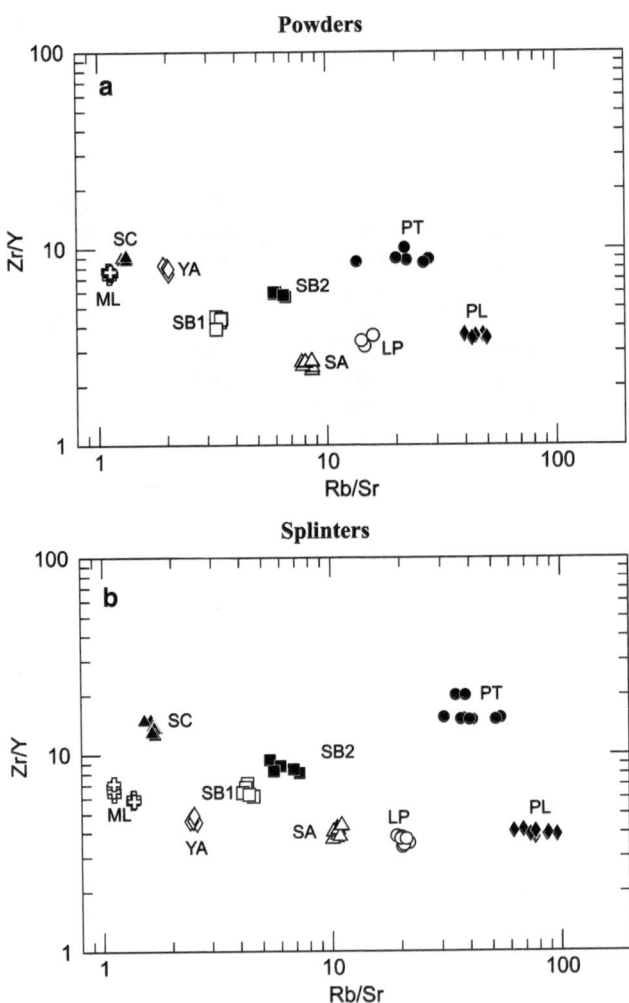

Fig. 5.5 A comparison between the destructive and non-destructive XRF methodologies. (**a**) Rb/Sr ppm vs. Zr/Y ppm on powders; (**b**) Rb/Sr vs. Zr/Y (X-rays intensity ratios) on entire splinters. Legend as in Fig. 5.3

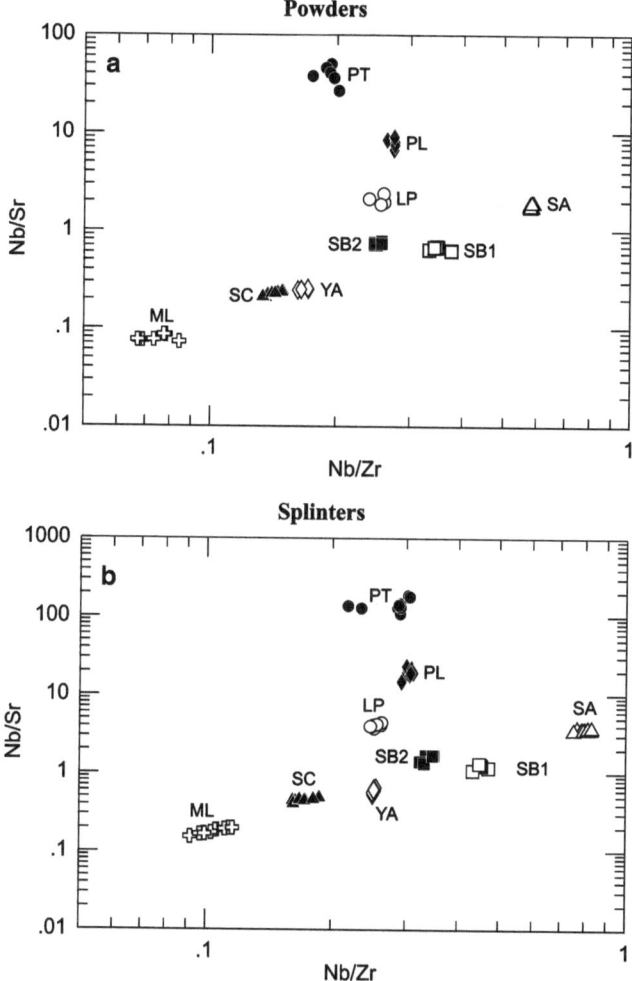

Fig. 5.6 A comparison between the destructive and non-destructive XRF methodologies. (**a**) Nb/Zr ppm vs. Nb/Sr ppm on powders; (**b**) Nb/Zr vs. Nb/Sr (X-rays intensity ratios) on entire splinters. Legend as in Fig. 5.3. Good correspondence exists between the different groupings, both for the ratio values and for the grouping geometry

obtained using the ratios between the absolute concentrations on powders of the same samples, as shown in Fig. 5.5a.

Similarly, the diagrams Nb/Zr vs. Nb/Sr in Figs. 5.6a and 5.6b show the ratios of the absolute concentrations and the intensity ratios, respectively, obtained with the two applied methodologies.

The comparison underlines, in an effective way, the similarity in the results obtained with the two methodologies and indicates the capacity of this non-destructive methodology in XRF to make a distinction between the obsidian sources in the Mediterranean area. It is important to note the high sensitivity of the non-destructive methodology even with a slight compositional difference in the field of the same obsidian source, observed mainly for Melos and Palmarola, along with the huge variation in Pantelleria sources. It is possible to use other combinations of analyzed chemical elements, but the few diagrams presented clearly show how it is possible to distinguish the principal sources of obsidian of the Mediterranean area.

The values of the concentration ratios never match the values of the intensity ratios exactly as data are obtained from two different methodologies; however, they show similar geometry proportions. Essentially, the observed differences are related to instrumental conditions, but they are not really significant for practical purposes and never influence the discriminating power of the groups.

With the application of the non-destructive methodology, the attribution of the origin of the archaeological artifacts is possible through the comparison between the intensity ratios obtained on sample splits of known origin and those on artifacts analyzed contemporaneously, and the splits of obsidians of known origin represent a further control over possible drifts of the spectrometer. In some cases, i.e., archaeological fragments of unsure provenance, it may be necessary to employ ternary or three-dimensional diagrams to attribute the provenance.

Archaeological Sites Investigated

In this chapter, we report the source area, obtained using the non-destructive WDXRF methodology, of about 1,400 obsidian fragments, mainly from Neolithic archaeological sites of Italy and Corsica. This methodology has also been used on archaeological fragments from European and South American archaeological sites (De Francesco et al. 2007; Biagi et al. 2007; Crisci et al. 1994). The obsidians analyzed have been selected by the archaeologists. In some cases, the origin of the entire group of artifacts discovered in a site has been determined, but in other cases only some representative samples have been selected after a macroscopic analysis. Thus, for several archaeological sites all the finds were analyzed, while for other sites a sample was selected.

The majority of the obsidians analyzed in the present work originate from the archaeological sites of Central Italy (Tuscany, Marches, Lazio, and Abruzzo), while the remainder are from sites located in Southern Italy (Campania and Apulia), Northern Italy (Emilia Romagna and Lombardy), Sardinia, and Corsica (France). The location of the archaeological Neolithic sites, along with the provenances of the analyzed obsidians, is shown in Fig. 5.7.

Listed below is contextual information on each of the main archaeological sites from where the analyzed obsidians originated. The number of the obsidians analyzed and their provenance are outlined in Tables 5.3–5.6, separated according to their geographical area.

Fig. 5.7 Location of the archaeological Neolithic sites from where the obsidians were analyzed and related provenances

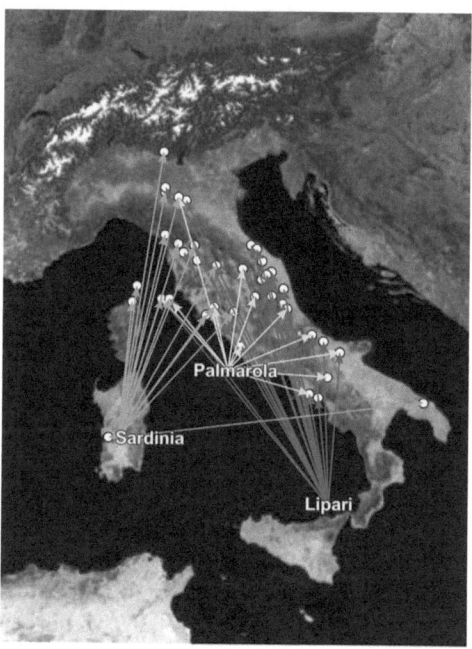

Central Italy

La Scola-Isola di Pianosa (Livorno)

The rock-cliff of La Scola is situated close to the eastern coast of Pianosa. The islet has an almost circular shape and a flat summit. On the northeastern side, there are a number of shelters under the rock that functioned in ancient times as real caves. With reference to the lithic industry, nearly 1,500 artifacts have been found, of which only 16% are tools. The most commonly used raw material is quartz (69%), followed by obsidian (20%), quartzite (7%), and flint (4%). All the materials were imported to the site (Radi and Danesi 2003; Ducci et al. 1999; De Francesco and Crisci 1999).

Cala Giovanna Piano: Isola Di Pianosa (Livorno)

The lithic industry discovered at this site is made up of 1,796 artifacts, of which nearly 90% are made of quartz and only 5% of obsidian. Altogether, the obsidian material is constituted by artifacts manufactured either as tools, flakes or blades, discards resulting from poor workmanship, and three nuclei. Due to the shortage of available artifacts, it is difficult to establish whether obsidian processing had been performed in the place, even when the presence of discards of workmanship and nuclei suggest this (Tozzi 2007; De Francesco and Crisci 1999; Bietti et al. 2004).

Grotta all'Onda (Lucca)

The first excavations made at Grotta all'Onda began in 1867 and were performed by the researchers of the Committee for the Search of Human Palaeontology of Florence, who reconstructed the principal stages of habitation in the cave. The most ancient prehistoric finds are datable to around 40,000 BP and is represented by a lithic industry with many flint tools. Small arrowheads and many small blades made of Sardinian obsidian have been recovered. These finds result from exchange activities when navigation was actively practiced. The contacts with Sardinia are also testified by the forms and decoration on ceramic artifacts (Campetti et al. 2001).

Marche Region

In the sixth millennium BC, the new "Neolithic" economy is characterized everywhere by the emergence of stable villages, breeding, agriculture, and pottery manufacture. In this period, the Marche region shows peculiar aspects, which differ between areas. The most ancient stage of the Marchigian Neolithic (settlement of Maddalena di Muccia) is datable within the sixth millennium BC in a calibrated chronology and shows, even in the types of artifacts, a difference in the latest moments of the same stage (settlement of Ripabianca di Monterado), which is datable between the end of the sixth and the beginning of the fifth millennium BC (Silvestrini 2003). A more advanced Neolithic stage in this region (settlements of Fontenoce di Recanati and Villa Panezia in Ascoli Piceno) can be placed in the first half of the fifth millennium BC. It is characterized by the existence of close relationships with the neighboring regions and with those that are relatively distant. These relationships are pointed out by the pottery that associates the typical forms of the advanced stage of the culture of Ripoli (Abruzzo) to the peculiar bends like Serra D'Alto e Diana (Silvestrini 2003). In the first half of the fourth millennium BC, the peopling of the Marche is mostly concentrated in the inside areas (e.g., settlements of Coppetella di Jesi, Donatelli di Genga, S. Maria in Selva di Treia) and along the valleys of the principal rivers (Misa, Esino, Potenza), while the occupation of the coastal areas is less frequent (Saline di Senigallia e Monte Tinello di Acquaviva Picena). The lithic industry of the Marchigian Neolithic also contains obsidian and smoothed greenstone (Antonelli et al. 2002, 2006).

Catignano (Pescara)

The site deposits have yielded 577 obsidian artifacts (8% of the lithic industry), dating between 6,300 and 5,900 BP. The raw material arrived in prepared cores, nevertheless the processing continues to be carried out on site, as testified by the high number of nuclei (De Francesco and Crisci 2003; Tozzi and Zamagni 2003).

Colle Cera (Pescara)

The Colle Cera site comprises a large Neolithic settlement, almost completely occupying the top of a hill at 237 m a.s.l. in the town of Loreto Aprutino (Abruzzo). The settlement belongs to the cultural setting known as *Cultura di Catignano – Scaloria Bassa,* a transitional phase between the ancient Neolithic (*Cultura della Ceramica Impressa*) and the middle Neolithic (Tozzi 1998, 2001; Tozzi and Zamagni 2003; De Francesco et al. 2008b; Barca et al. 2008).

La Marmotta: Anguillara (Roma)

The excavations started in 1989 on the floor of Lake Bracciano in the place called "La Marmotta," near Anguillara Sabazia (Rome). They added new information on the reconstruction of the life style of a Neolithic village. The calibrated carbon dating places the use of the site between 5750 and 5260 BC and qualifies the "La Marmotta" village as the most ancient Neolithic bank settlement of western Europe known at present. Particularly abundant is the lithic industry expressed in stone, flint, and obsidian (Fugazzola Delfino et al. 2004).

The results on the provenance of the obsidians from the sites of Central Italy are shown in Table 5.3.

Northern Italy

Bazzarola (Reggio Emilia)

The 2003 excavations yielded stratigraphical data of national importance, which are useful to outline some of the scientific problems related to the agricultural colonization of the Pianura Padana. The first farmers arrived in Emilia from the coast. The lithic industry is entirely made of lithotypes coming from the Appennine (flint pebbles). The Lessinian flints are missing, while the obsidian discards of workmanship are frequent and after this first appearance in the Pianura Padana, the other discards found are only from the Middle Neolithic.

Benefizio (Parma)

The excavations performed in the rotunda of the street named Via Spezia follow those already made in the near area of Benefizio, which at the end of 2002 brought to light interesting finds of the Copper Age (fourth and third millennium BC). Finds of the same type have also been recovered in the subway of Via Spezia. At a depth of about 3 m under the ground, artifacts of Neolithic age have been discovered (fourth millennium BC), which reveal the identity of the first farmers of Parma.

Table 5.3 Results of the provenance of the obsidians from the sites of Central Italy

Number	Site	Lipari	Palmarola	Pantelleria	Sardinia	Unknown	Overall samples
1	Botteghino Pontedera (PI)	1	0	0	0	0	1
2	Cala Giovanna (LI)	1	9	0	162	4	176
3	Cala Giovanna Piano (LI)	2	1	0	38	1	42
4	Casa dell'Isola (LU)	0	0	0	6	0	6
5	Castiglione in Tever- ina (VT)	0	0	0	1	0	1
6	Catignano (PE)	168	13	0	0	2	183
7	Chiarentana (SI)	17	5	0	1	0	23
8	Colle Santo Stefano (AQ)	0	24	0	0	0	24
9	Collecera	29	4	0	0	2	35
10	Coppetella di Jesi (AN)	5	0	0	0	0	5
11	Donatelli di Genga (AN)	2	0	0	0	0	2
12	Fontenoce di Recana- ti (MC)	1	0	0	0	0	1
13	Gotta della Spinosa (GR)	0	0	0	6	0	6
14	Grotta all'Onda (LU)	0	0	0	34	1	35
15	La Scola (LI)	1	7	0	20	1	29
16	Maddalena di Muccia (MC)	6	2	0	0	0	8
17	Marche Mar 02	22	3	0	1	1	27
18	Marmotta (Roma)	55	338	0	0	12	405
19	Monte Frignone (LU)	0	0	0	0	2	2
20	M.te Tinello di Ac- quaviva Picena (AP)	1	0	0	0	0	1
21	Mulino	1	0	0	0	0	1
22	Neto Via Verga (FI)	21	0	0	4	0	25
23	Ripabianca di Mon- terado (AN)	5	0	0	0	0	5
24	Ripoli Fossecesia (CH)	36	5	1	0	2	44
25	Saline di Senigallia (AN)	2	0	0	0	0	2
26	Santa Maria in Selva (AN)	31	0	0	0	1	32
27	Settefonti (AQ)	3	6	0	0	1	10
28	Spazzavento (FI)	22	0	0	1	8	31
29	Villa Panezia di As- coli (AP)	0	1	0	0	0	1
	Overall Central Italy	*432*	*418*	*1*	*274*	*38*	*1163*

Table 5.4 Results of the provenance of the obsidians from the sites of Northern Italy

Number	Site	Lipari	Palmarola	Pantelleria	Sardinia	Unknown	Overall samples
1	Bazzarola (RE)	5	0	0	9	0	14
2	Benefizio (PR)	2	0	0	0	0	2
3	Gaione (PR)	4	0	0	0	0	4
4	Le Mose (PC)	2	0	0	0	0	2
5	Monte Covolo (BS)	0	0	0	5	0	5
6	Ponteraro (PR)	10	8	0	0	2	20
7	Travo (PC)	1	0	0	4	2	7
	Overall Northern Italy	*24*	*8*	*0*	*18*	*4*	*54*

Pieces of "Appenninic flint" and crocks of "engraved ceramics" show that those who brought the agriculture to Parma originated from the Ligurian coasts.

The results of the provenance of the obsidians from the sites of Northern Italy are shown in Table 5.4.

Southern Italy

Botteghelle (Napoli)

During the archaeological investigations in the Napolitan urban territory, close to Viadotto Botteghelle, a site rich in archaeological evidence has been discovered. In addition to the remains of a Neolithic village in the deepest layers of the ground, finds from the fourth millennium BC have been recovered, while in more superficial layers, numerous materials have emerged that lead us to hypothesize that a sanctuary was present. Diagnostic artifacts suggest a date between the end of the fourth and the second centuries BC.

Oria Sant'Anna (Taranto)

The first traces of a Neolithic population in this area are dated to around the fifth millennium BC. The most meaningful peopling took place only during the late Neolithic, as shown by the notable concentration of settlements on the hills of Sant'Anna. These settlements were located in Monte Papalucio, where the traces of not a regular occupation have been found but constituted only by few lithic fragments. This pattern occurs in the region of Contrada Pappadà, close to San Giovanni lo Parete, with the recovery of obsidian tools; in the region of Contrada Monti, with trapezoidal blades and again obsidian tools; and in the region of Contrada Fontane, where trapezoidal blades and obsidian blades were found. Traces of a Neolithic occupation have also been discovered in the place called

Table 5.5 Results of the provenance of the obsidians from the sites of Southern Italy

Number	Site	Lipari	Palmarola	Pantelleria	Sardinia	Unknown	Overall samples
1	Ariano Irpino (AV)	0	1	0	0	0	1
2	Botteghelle (NA)	4	1	0	0	0	5
3	Grotta della Serratura (SA)	5	0	0	0	0	5
4	Masseria di Gioia (BR)	0	0	0	12	0	12
5	Oria Sant'Anna (TA)	36	1	0	0	0	37
6	Ripatetta (FG)	1	0	0	0	0	1
	Overall Southern Italy	*46*	*3*	*0*	*12*	*0*	*61*

Madonna della Scala, of the Canale Reale, where blade fragments and obsidian blades have been found.

The results of the provenance of the obsidians from the sites of Southern Italy are shown in Table 5.5.

Sardinia

Torre Foghe (Oristano)

The site is located on the western coast of Sardinia at the mouth of Rio Mannu. The lithic industry consists of rocks, belonging to the fonolite group (10%), flint originated by hydrothermal processes (10%), and obsidian (80%). The obsidian is constituted by 938 artifacts and 107 tools (De Francesco and Bocci 2007; Dini 2007).

Corsica

Lumaca

The site is situated at a height of 450 m, on the crest of a hill in the northwest extremity of Cap Corse. It is constituted of several adjoining terraces. A series of soil corings performed on two terraces revealed several levels of occupation starting from the early Neolithic. The lithic materials discovered originate from drill holes 2, 3, and 4. Sixty-seven obsidian fragments originated from drill hole 3. Obsidian constitutes 60% of all the lithic materials discovered, thus making it the most used lithic type. From a macroscopic analysis of the fragments, four varieties of obsidian have been recognized: (1) opaque black, more or less bright, ($n = 25$); (2) translucent and sometimes with a smoky gray aspect ($n = 15$); (3) opaque and

Table 5.6 Results of the provenance of the obsidians from the sites of Sardinia and Corsica

Number	Site	Lipari	Palmarola	Pantelleria	Sardinia	Unknown	Overall templates
1	Costa di U Monte (Corsica)	0	0	0	50	0	50
2	Lumaca (Corsica)	0	0	0	21	0	21
3	Torre Foghe Tres- nuraghes (OR)	0	0	0	49	0	49
	Overall Sardinia and Corsica	*0*	*0*	*0*	*120*	*0*	*120*

with whitish parallel striations ($n = 5$); (4) transparent and bright (Lorenzi 1999). The analysis was conducted only on 21 fragments (De Francesco and Crisi 1999; De Francesco 2002a).

Costa Di U Monte

The site is situated on the east coast of Corsica, on a hill with a flat summit at a height of 213 m and 2 km from the coast. Occupational stages are known to exist during and after the Neolithic period. Approximately 800 obsidian elements have been found (Marini et al. 2007).

The results on the provenance of the obsidians from the sites of Sardinia and Corsica are shown in Table 5.6.

Results

The application of the non-destructive WDXRF methodology to the archaeological obsidian fragments from numerous Italian Neolithic sites and from the island of Corsica (France) allowed the attribution of 1,356 samples out of 1,398.

The results obtained show that for the archaeological sites of Central Italy, the majority of obsidian originates from Sardinia, Lipari, and Palmarola. In Northern Italy, the majority of obsidians derive from Lipari and from Monte Arci in Sardinia. In Southern Italy, obsidian provenance from Lipari, Sardinia, and Palmarola prevail. Only Sardinian origins have been found for the sites of Lumaca and Costa di U Monte in Corsica and Torre Foghe in Sardinia, as shown in Fig. 5.8.

Of the entire group of archaeological obsidians analyzed, 36% come from Lipari, 31% from Palmarola, and 30% from Sardinia (Fig. 5.9). It is noteworthy that the remaining percentage of the unassigned archaeological obsidians is very low (around 3%), and in most cases this can be attributed to the extremely small size (below a centimeter in diameter) or the very slight thickness of the fragments.

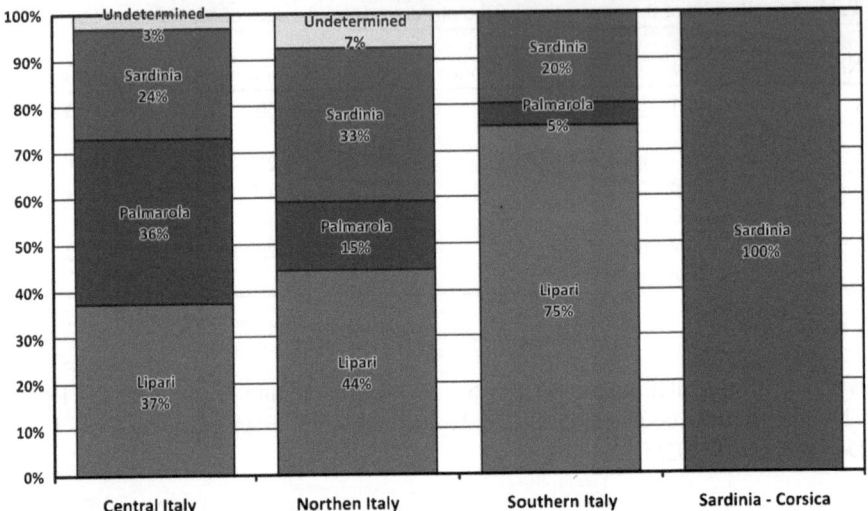

Fig. 5.8 Archaeological obsidian provenance, based upon the geographical areas

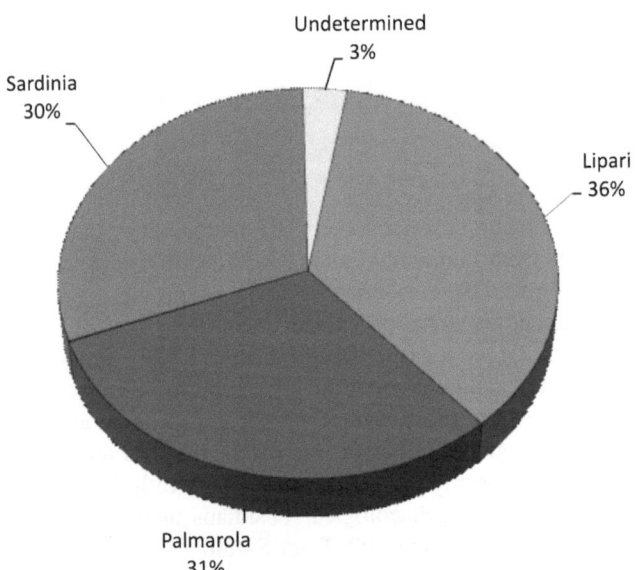

Fig. 5.9 Overall pie diagram of the provenances for 1,398 archaeological samples analyzed, the provenance of the 3% alone was not attributable

Conclusions

Numerous samples of obsidians, collected from the principal geological outcrops of archaeological interest in the Mediterranean area (Lipari, Pantelleria, Sardinia, Palmarola, and the Greek islands of Melos and Gyali), have been analyzed with the WDXRF routine on powders (destructive) and with the non-destructive WDXRF methodology on entire fragments of the same samples. With the non-destructive XRF methodology, only the secondary X-ray intensities of five trace elements (Nb, Sr, Zr, Rb and Y) have been determined. These trace elements are particularly indicative of genetic processes and, therefore, allow us to discriminate all the obsidian sources in the Mediterranean area with high precision.

The comparison between the results obtained by using the two methodologies shows that the X-ray intensity ratios obtained on samples of geological obsidians are able to effectively discriminate between the obsidian sources just as accurately as with concentrations on powders.

The non-destructive WDXRF analyses, conducted on 1,398 entire archaeological artifacts of unknown origin from numerous Italian Neolithic sites and from the island of Corsica (France), allow us to trace the origin of 1,356 artifacts, with the same precision and reliability as that obtained with the destructive methodology, reducing cost and analysis time, when compared with other more sophisticated methodologies. The results obtained by LA-ICP-MS method on the geological obsidian of the Mediterranean area (Barca et al. 2007) and on the artifacts from the Colle Cera archaeological site (Barca et al. 2008) confirm the numerous advantages of the non-destructive WDXRF methodology, taking into consideration the small number of unassigned archaeological obsidians (around 3%), and, principally, the very small dimensions of the archaeological samples not completely irradiated with X-rays.

References

Acquafredda, P., Adriani, T., Lorenzoni, S., Zanettin, E. (1996). Proposal of a non destructive analytical method by SEM-EDS to discriminate Mediterranean obsidian sources. *Advances in Clay Minerals*, 269–271.

Acquafredda, P., Adriani, T., Lorenzoni, S., Zanettin, E. (1999). Chemical characterization of obsidian from different Mediterranean sources by non destructive SEM-EDS analytical method. *Journal of Archaeological Science*, 26, 315–325.

Antonelli, F., Crisci, G.M., De Francesco, A.M., Silvestrini, M. (2002). Provenienza delle ossidiane di siti archeologici neolitici delle Marche. *PLINIUS, supplemento italiano all' European Journal of Mineralogy*, 28, 28–29.

Antonelli, F., Crisci, G.M., De Francesco, A.M., Silvestrini, M. (2006). Le ossidiane degli insediamenti neolitici delle Marche: provenienza ed implicazioni. *Proceedings of the V Convegno di Archaeometria "Innovazioni Tecnologiche per i Beni Culturali in Italia"* (pp. 301–309). Bologna: Patron Editore.

Barberi, F., Borsi, S., Ferrara, G., Innocenti, F. (1967). Contributo alla conoscenza vulcanologia e magmatologica delle isole dell'arcipelago Pontino. *Memorie della Società Geologica Italiana*, 17, 581–606.

Barca, D., De Francesco, A.M., Crisci, G.M. (2007). Application of Laser Ablation ICP–MS for characterization of obsidian fragments from peri-Tyrrhenian area. *Journal of Cultural Heritage*, 8, 141–150.

Barca, D., De Francesco, A.M., Crisci, G.M., Tozzi C. (2008). Provenance of obsidian artifacts from site of Colle Cera, Italy, by LA-ICP-MS method. *Periodico di Mineralogia*, 77, 39–50.

Bellot-Gurlet, L., Le Bourdonnec, F., Popeau, G., Dubernet, S. (2004). Raman micro-spectroscopy of western Mediterranean obsidian glass: one step towards provenance study? *Journal of Raman Spectroscopy*, 35, 671–677.

Biagi, P., De Francesco, A.M., Bocci, M. (2007). New data on the archaeological obsidian from the middle-late Neolithic and chalcolithic sites of the Banat and Transylvania. In: Kozowski, J.K., Raczky, P. (Eds) *The Lengyel, Polgar and Related Cultures in the Middle/Late Neolithic in Central Europe* (pp. 309–326). Krakow: Jagiellonian University and ELTE.

Bietti, A., Boschian, G., Crisci, G.M., Danese, E., De Francesco, A.M., Dini, M., Fontana, F., Giampietri, A., Grifoni, R., Guerreschi, A., Liagre, J., Negrino, F., Radi, G., Tozzi, C., Tykot R. (2004). Inorganic raw materials economy and provenence of chipped industry in some stone age sites of Northern and Central Italy. *Collegium Antropologicum*, 28(1), 41–54.

Bigazzi, G., Bonadonna, F.P. (1973). Fission track dating of the obsidian of Lipari Island (Italy). *Nature*, 242, 322–323.

Bigazzi, G., Radi, G. (1981). Datazioni con le tracce di fissione per l'identificazione della provenienza dei manufatti in ossidiana. *Rivista di Scienze Preistoriche*, 36, 223–250.

Bigazzi, G., Bonadonna, F.P., Belluomini, G., Malpieri, L. (1971). Studi sulle ossidiane italiane. IV. Datazione con il metodo delle tracce di fissione. *Bollettino della Societa' Geologica Italiana*, 90, 469–480.

Bigazzi, G., Yegingil, Z., Ercan, T., Oddone, M., Ozdogan, M. (1993). Fission track dating obsidian in Central and Northern Anatolia. *Bulletin of Volcanology*, 55, 588–595.

Birò, T.K., Pozsgai, I., Vlader, A. (1986). Electron beam microanalyses of obsidian samples from geological and archaeological sites. *Acta Archaeologica Academiae Scientiarum Hungaricae*, 38, 257–258.

Campetti, S., Dodaro, S., Ferrini, G., Crisci, G.M., De Francesco, A.M., Montanari, C., Guido, M., Cozzani, M., Perrini, L., Berton, A., Bigini, I., Turini, R. (2001). Risultati preliminari di nuove indagini nel deposito preistorico di Grotta all'Onda, Camaiore (Lucca). *Atti della XXXIV Riunione Scientifica dell'Istituto di Storia e Protostoria*, 349–366.

Cann, J.R., Renfrew, C. (1964). The characterisation of obsidian and its application to the Mediterranean region. *Proceeding of the Prehistoric Society*, 30, 111–133.

Cirrincione, R., Di Martino, G., Lo Giudice, A., Pappalardo, G., Pappalardo, L., Pezzino, A. (1995). Analisi non distruttiva di Ca e Sr eseguita mediante un sistema XRF portatile su diversi tipi di calcare provenienti da cave di Palazzolo e Monti Climiti. *Proceeding of 1st international Congress on: "Science and Technology for the Safeguard of Cultural Heritage in the Mediterranean Basin"* 27 Nov–2 Dec, Catania, Siracusa – Italy.

Civetta, L., Cornette, Y., Crisci, G.M., Gillot, P.Y., Orsi, G., Requeios, C.S. (1984). Geology, geochronology and chemical evolution of the island of Pantelleria. *Geological Magazine*, 121 (6), 541–668.

Civetta, L., Cornette, Y., Gillot, P.Y., Orsi, G. (1988). The eruptive history of Pantelleria (Sicily Channel). *Bulletin of Volcanology*, 50, 47–57.

Cornette, Y., Crisci, G.M., Gillot, P.Y., Orsi, G. (1983). Recent volcanic history of Pantelleria: a new interpretation. *Journal of Volcanology and Geothermal Research*, 17, 361–373.

Crisci, G.M., De Rosa, R., Esperança, S., Mazzuoli, R., Sonnino, M. (1991). Temporal evolution of a three component system: the island of Lipari (Aeolian arc, Southern Italy). *Bulletin of Volcanology*, 53, 207–221.

Crisci, G.M., Ricq-De Bouard, M., Lanzafame, U., De Francesco, A.M. (1994). Nouvelle méthode d'analyse et provenance de l'ensemble des obsidiennes neolithiques du Midi de la France. *Gallia Préistoire*, 36, 299–327.

De Francesco, A.M., Bocci, M. (2007). Risultati dell'analisi non distruttiva in XRF sulle ossidiane dei siti archaeologici di Cala Giovanna Piano (Arcipelago Toscano) e Torre Foghe. In: a cura di Tozzi, C., Weiss, M.C. *Preistoria e Protostoria dell'area Tirrenica* (121 p.). Pisa: Ed. Felici.

De Francesco, A.M., Crisci, G.M. (1999). Provenienza delle ossidiane dei siti archaeologici di Pianosa (Arcipelago Toscano) e Lumaca (Corsica) con il metodo non distruttivo in Fluorescenza X. In: Tozzi, C., Weiss, M.C. (Eds) *Il primo popolamento Olocenico dell'area corso-toscana* (pp. 253–258). Pisa: Edizioni. ETS.

De Francesco, A.M., Crisci, G.M. (2003). L'ossidiana. In: Tozzi, C., Zamagni, B. (Eds) *Gli scavi nel villaggio neolitico di Catignano (1971–1980)* (pp. 239–240). Firenze: Origines, Studi e materiali pubblicati a cura dell'Istituto Italiano di Preistoria e Protostoria.

De Francesco, A.M., Crisci, G.M., Lanzafame U. (2000). Non-destructive analytical method by XRF for provenancing archaeological obsidians from Mediterranean Area. *Bollettino della Accademia Gioenia di Scienze Naturali*, 357, 229–239.

De Francesco, A.M., Crisci, G.M., Radi, G., Tozzi, C., Perazzi, P., Ducci, S. (2002a). Provenance of the obsidians coming from Catignano (Abruzzo), Pianosa (Tuscany Archipelago) and Lumaca (Corsica) neolithic sites with non-destructive X-ray Fluorescence analytical method. *2° Congresso Nazionale Scienza e Beni Culturali*, 29 gennaio–1 febbraio, Bologna.

De Francesco, A.M., Crisci, G.M., Bocci, M., Lanzafame, U. (2002b). Provenienza delle ossidiane archeologiche di alcuni siti neolitici italiani. *PLINIUS, supplemento italiano all'European Journal of Mineralogy*, 28, 137–138.

De Francesco, A.M., Bocci, M., Crisci, G.M. (2004). Provenienza dal Monte Arci per le ossidiane di alcuni siti archeologi dell'Italia Centrale e della Francia Meridionale. *Proceeding of 2° Convegno Internazionale "L'ossidiana del Monte Arci nel Mediterraneo* (pp. 303–309), Pau (Cagliari) 28–30 November. Ed. AV (CA).

De Francesco, A.M., Bocci, M., Crisci, G.M., Lanzafame, U. (2005). Caratterizzazione archeometrica delle ossidiane del Monte Arci: confronto fra metodologia tradizionale in XRF e metodologia XRF non distruttiva. *Proceeding of III° Convegno Internazionale "L'ossidiana del Monte Arci nel Mediterraneo"* (pp. 117–128). Ed. PTM (OR).

De Francesco, A.M., Bocci, M., Crisci, G.M., Martini, F., Tozzi, C., Radi, G., Sarti, L., Cuda, M.T. (2006). Applicazione della metodologia analitica non distruttiva in Fluorescenza X per la determinazione della provenienza delle ossidiane archeologiche del progetto "Materie Prime" dell'I.I.P.P. *Atti della XXXIX Riunione Scientifica dell'Istituto di Storia e Protostoria.*

De Francesco, A.M., Durán, V., Bloise, A., Neme, G. (2007). Caracterización Y Procedencia de obsidianas de Sitios Arqueológicos del Area Natural Protegida Laguna del Diamante (Mendoza, Argentina) con Metodología no Destructiva por Fluorescencia de Rayos (Xrf). In: "Arqueología y Ambiente de Areas Naturales Protegidas de la Provincia de Mendoza". Víctor Durán y Valeria Cortegoso editores. *Número especial de la Revista Anales de Arqueología y Etnología, Volumen 61*. FFyl. Mendoza, Argentina: Universidad Nacional de Cuyo.

De Francesco, A.M., Crisci, G.M., Bocci, M. (2008a). Non-destructive analytical method by XRF for determination of provenance of archaeological obsidians from the Mediterranean Area. A comparison with traditional XRF method. *Archaeometry*, 50(2), 337–350.

De Francesco, A.M., Francaviglia, V., Bocci, M., Crisci, G.M. (2008b). Archaeological obsidians provenance of several Italian sites using non destructive XRF Method. *International Specialized Workshop – The Dating and Provenance of Obsidian and Ancient Manufactured Glass*. Delphi, Greece, 21–24 February.

Di Paola, G.M. (1974). Volcanology and petrology of Nysiros Island (Dodecanese, Grece). *Bulletin de Volcanologie*, 38, 944–987.

Dini, M. (2007). L'industria in ossidiana dei siti neolitici di S. Caterina Pittinuri e Torre Foghe sulla costa occidentale della Sardegna. In: a cura di Tozzi, C., Weiss, M.C. *Preistoria e Protostoria dell'area Tirrenica* (pp. 195–202). Pisa: Ed. Felici.

Ducci, S., Guerrini, M.V., Perazzi, P. (1999). L'insediamento de La Scola (Isola di Pianosa, Comune di Campo nell'Elba, LI). In: Tozzi, C., Weiss, M.C. (Eds) *Il primo popolamento Olocenico dell'area corso-toscana* (pp. 83–90). Pisa: Edizioni. ETS.

Francaviglia, V. (1984). Characterization of Mediterranean obsidian souces by classical petro-chemical methods. *Preistoria Alpina – Museo Tridentino di Scienze Naturali di Trento (Italy)*, 20, 311–332.

Francaviglia, V. (1988). Ancient obsidian sources on Pantelleria (Italy). *Journal of Archaeological Science*, 15, 109–122.

Franzini, M., Leoni, L., Saitta, M. (1972). A simple method to evaluate the matrix effects in X-ray fluorescence analysis. *X-ray Spectrometry*, 1, 151–154.

Franzini, M., Leoni, L., Saitta, M. (1975). Revisione di una metodologia analitica per fluorescenza X basata sulla correzione completa degli effetti di matrice. *Rendiconti della Società Italiana di Mineralogia e Petrologia*, 31, 365–378.

Fugazzola Delfino, M.A., Pessina, A., Tiné, V. (2004). Il Neolitico in Italia. Ricognizione, catalogazione e pubblicazione dei dati bibliografici, archivistici, materiali e monumentali. Volume III – Siti a cura di Maria Antonietta Fugazzola Delpino, Andrea Pessina e Vincenzo Tiné. Firenze: Istituto Italiano di Preistoria e Protostoria, Roma, Soprintendenza al Museo Preistorico Etnografico "L. Pigorini", 453 p.

Fyticas, M., Giuliani, O., Innocenti, F., Marinelli, G., Mazzuoli, R. (1976). Geochonological data on recent magmatism of Aegean Sea. *Tectonophysics*, 31, T29–T34.

Fyticas, M., Innocenti, F., Kolios, N., Manetti, P., Mazzuoli, R., Poli, G., Rita, F., Villari, L. (1986). Volcanology and petrology of volcanic products from the Island of Milos and neigh bourig islets. *Journal of Volcanology and Geothermal Research*, 28, 297–317.

Glascock, M.D., Neff, H. (2003). Neutron activation analysis and provenance research in archae-ology. *Measurement Science and Technology*, 14, 1516–1526.

Gratuze, B. (1999). Obsidian characterization by laser ablation ICP-MS and its application to prehistoric trade in the Mediterranean and the near east: sources and distribution of obsidian within the Aegean and Anatolia. *Journal of Archaeological Science*, 26, 869–881.

Hallam, B.R., Warren, S.E., Renfrew, C. (1976). Obsidian in the western Mediterranean: charac-terization by neutron activation analysis and optical emission spectroscopy. *Proceeding of the Prehistoric Society*, 42, 85–110.

Kayani, P.I., McDonnel, G. (1996). An assessment of back-scattered electron petrography as a method for distinguishing Mediterranean obsidians. *Archaeometry*, 38, 43–58.

Langella, M., Boscaino, M., Coubrai, S., Curci, A., De Francesco, A.M., Senatore, M.R. (2003). Baselice (Benevento): Il sito pluristratificato neolitico di torrente Cervaro. *Rivista di Scienze Preistoriche*, LIII, 259–336.

Le Bas, M.J., Le Maitre, R.W., Streckeisen, A., Zanettin, R. (1986). A chemical classification of volcanic rocks based on the total alkali-silica diagram. *Journal of Petrology*, 27/3, 745–750.

Leoni, L., Saitta, M. (1976). X-ray fluorescence analysis of 29 trace elements in rocks and mineral standards. *Rendiconti della Società Italiana di Mineralogia e Petrologia*, 32, 497–510.

Lorenzi, F. (1999). Lumaca: un example de neolithisation au nord du Cap Corse. In: Tozzi, C., Weiss, M.C. (Eds) *Il primo popolamento Olocenico dell'area corso-toscana* (pp. 133–138). Pisa: Edizioni. ETS.

Machey, M.P., Warren, S.E. (1983). The identification of obsidian source in the M. Arci region of Sardinia. In: Aspinall, A., Warren, S.E. (Eds) *Proceedings of the 22nd Symposium on Archaeometry* (pp. 420–431). Bradford: University of Bradford.

Marini, N., De Francesco, A.M., Bocci, M., Bressy, C., Gratuze, B. (2007). Costa di U Monte – du Néolithique à l'âge du Fer sur la cote orientale corse: résultats de fouilles et provenance des vestiges. In: a cura di Tozzi, C., Weiss, M.C. *Préhistoire et protohistoire de l'aire tyrrhénienne* (pp. 35–42). *PROGETTO INTERREG III A Francia – Italia -Isole Toscana, Corsica, Toscana, Sardegna RICERCA 3.1. Unione Europe*a. Pisa: Ed. Felici.

Montanini, A., Villa, J.M. (1993). $^{40}Ar/^{39}Ar$ chronostratigraphy of Monte Arci volcanic complex (Western Sardinia, Italy). *Acta Volcanologica*, 3, 229–233.

Montanini, A., Barbieri, M., Castorina, F. (1994). The role of fractional crystallization, crustal melting and magma mixing in the petrogenesis of rhyolites and mafic inclusion-bearing dacites from the Monte Arci volcanic complex (Sardinia, Italy). *Journal of Volcanology and Geothermal Research*, 61, 95–120.

Radi, G., Danesi, E. (2003). Il sito neolitico di Settefonti a Prata d'Ansidonia (AQ). *Atti della XXXVI Riunione Scientifica dell'Istituto di Storia e Protostoria*, 163–179.

Renfrew, C., Cann, J.R., Dixon, J.E. (1965). Obsidian in the Aegean. *Annual of the British School Athens*, 60, 225–247.

Shackley, M.S. (Ed). (1998a). *Archaeological Obsidian Studies: Method and Theory*. New York: Springer/Plenum.

Shackley, M.S. (1998b). Gamma rays, X-rays, and stone tools: some current advances in archaeological geochemistry. *Journal of Archaeological Science*, 25, 259–270.

Shackley, M.S. (2002). Precision versus accuracy in the XRF analysis of archaeological obsidian: some lesson for archaeometry and archaeology. In: Jerem, E., Biro, K.T. (Eds) *Proceedings of the 31st Symposium on Archaeometry* (pp. 805–810), Budapest, Hungary. Oxford: BAR International Series 1043(II).

Shackley, M.S. (2005). *Obsidian: Geology and Archaeology in the North American Southwest*. Tucson: University of Arizona Press.

Silvestrini, M. (2003). Dal Paleolitico all'età del Bronzo. In: a cura di Luni M. *Archeologia nelle Marche* (pp. 21–30). Firenze: Nardini Editore.

Stewart, S.J., Cernicchiaro, G., Scorzelli, R.B., Poupeau, G., Acquafredda, P., De Francesco, A.M. (2003). Magnetic properties and 57 Fe Mossbauer spectroscopy of Mediterranean prehistoric obsidians for provenance studies. *Journal of Non-Crystalline Solids*, 323, 188–192.

Summerhayes, G., Bird, J.R., Fullagar, R., Gosden, C., Specht, J., Torrence, R. (1998). Application of PIXE-PIGME to obsidian characterization on west New Britain, Papua New Guinea. In: Shackley, M.S. (Ed) *Archaeological Obsidian Studies: Method and Theory*. New York: Springer/Plenum.

Thorpe, W.O. (1995). Obsidian in the Mediterranean and the Near East: a provenancing success story. *Archaeometry*, 37, 217–248.

Thorpe, W.O., Warren, S.E., Courtin, J. (1984). The distribution and sources of archaeological obsidian from Southern France. *Journal of Archaeological Science*, 11, 135–146.

Tozzi, C. (1998). Culture de Catignano-Scaloria Bassa. In: Guilaine, J. (Ed) *Atlas du Néolithique européen, vol. 2A, L'Europe occidentale, ERAUL 46, Etudes et Recherches de l'Université de Liège* (pp. 178–180).

Tozzi, C. (2001). Ripa Tetta e Catignano, établissements néolithiques de l'Italie adriatique, in "Communautés villageoises du Proche Orient à l'Atlantique", Séminaire du Collège de France, sous la direction de J. Guilaine, Ed. Errance, pp. 153–167.

Tozzi, C. (2007). Considerazioni riassuntive sull'insediamento di Cala Giovanna – Piano. In: a cura di Tozzi C., Weiss M.C. *Preistoria e Protostoria dell'area Tirrenica* (pp. 167–168). Pisa: Ed. Felici.

Tozzi, C., Zamagni, B. (2003). *Gli scavi nel villaggio neolitico di Catignano (1971–1980)* (pp. 239–240). Firenze : Origines, Studi e materiali pubblicati a cura dell'Istituto Italiano di Preistoria e Protostoria.

Tykot, R.H. (1997). Characterization of the Monte Arci (Sardinia) Obsidian Sources. *Journal of Archaeological Science*, 24, 467–479.

Tykot, R.H. (2002). Chemical fingerprinting and source-tracing of obsidian: The central Mediterranean trade in black gold. *Accounts of Chemical Research*, 35, 618–627.

Tykot, R.H., Young, S.M.M. (1996). Archaeological applications of inductively coupled plasma-mass spectrometry. In: Orna, M.V. (Ed) *Archaeological Chemistry*, ACS Symposium series 625 (pp. 116–130). Washington, DC: American Chemical Society.

Chapter 6
Portable XRF of Archaeological Artifacts: Current Research, Potentials and Limitations

Ioannis Liritzis and Nikolaos Zacharias

Introduction

Nuclear beams and lasers are becoming increasingly important as analytical tools in art and archaeology for dating and characterization studies. Their portability and multifunctional mode, however, is the main concern as it requires particular instrumentation and software development. This development is triggered by the need to study the materials of cultural heritage, whether in situ or in the museums, in a non-invasive, rapid manner so as to acquire the maximum possible information in one operational action. The portable X-ray fluorescence (PXRF) instrumentation is aligned along this leading research trend.

For rapid, non-destructive detection of chemical elements, portable or desktop XRF devices are available, either in normal mains power or battery-operated. They are rugged, and highly sensitive for performing in-the-field non-destructive elemental analysis. Such system designs are compact, low weight, consume less power, and can operate at varying voltages of 40 to 60 kV and variable currents in the range of microamps, obtaining accurate results quickly.

This XRF technology can generally detect amounts of major, minor, and trace chemical elements. In particular, it is uniquely capable of detecting trace amounts of heavy elements, such as barium (Ba), antimony (Sb), lead (Pb), and strontium (Sr). The technology can be used wherever rapid, non-destructive, in situ analysis of chemical elements is needed.

To rapidly determine the identity and quantity of elements and distinguish them from background radiation, the analysis process can divide the spectrum into energy sub-bands, reducing the algorithmic complexity of analyzing the entire spectrum at once. As a result, the system obtains accurate, virtually real-time results

I. Liritzis (✉)
Department of Mediterranean Studies, Laboratory of Archaeometry, University of the Aegean, Rhodes 85100, Greece
e-mail: liritzis@rhodes.aegean.gr

M.S. Shackley (ed.), *X-Ray Fluorescence Spectrometry (XRF) in Geoarchaeology*,
DOI 10.1007/978-1-4419-6886-9_6, © Springer Science+Business Media, LLC 2011

for hundreds of material classes. These algorithms also could be applied to neutron and gamma spectroscopic analysis. Such a system with a low-flux, micro-focused X-ray tube operates using a high-voltage, low-power, miniaturized, and regulated power supply.

These apparatuses are developed for many investigations, including archaeology and culture materials culture in general (Potts and West 2008). The benefits encompass: *Sensitivity*: This technology can detect materials at the parts-per-million (ppm) level. *Advanced analytics*: Using software algorithms, the technology detects the relative amounts of elemental constituents, which can be used to identify the possible presence of surface traces. *Automated, remote operations*: They can be fully automated and operated remotely if required.

Especially, the systems designed with an X-ray tube that can operate up to 60 keV provide the ability to complete measurements to the ppm-level in a shorter time and to identify heavier elements more easily than with the 40 keV tubes typically found in portable XRF systems. The cost of a portable XRF systems that uses radioisotopes or X-ray generators ranges between 15,000 and 60,000 €s ($20,000–$80,000).

The non-destructive advantage of the PXRF should be emphasized parallelly with their portability because other non-destructive methods of analysis are available with even superior capabilities that measure a large number of elements (e.g., particle induced X-ray emission [PIXE], NAA, IRPAS, FTIR, SIMS), but they are not portable. The portability coupled with the non-invasive capability makes the PXRF systems more favored by archaeometrists and, especially, archaeological scientists.

XRF spectrometry typically uses a polychromatic beam produced from radio-isotopes X-ray tubes or synchrotrons of short wavelength/high-energy photons to induce the emission of longer wavelength/lower energy characteristic lines in the sample to be analyzed (see Chap. 2). X-ray spectrometers may use either the diffracting power of a single crystal to isolate narrow wavelength bands (wavelength-dispersive XRF [WDXRF]) or an energy-selective detector to isolate narrow energy bands (energy-dispersive XRF [EDXRF]) from the polychromatic radiation (including characteristic radiation) that is produced in the sample. Because the relationship between emission wavelength and atomic number is known, isolation of individual characteristic lines allows the unique identification of an element to be made, and elemental concentrations can be estimated from characteristic line intensities. WDXRF instrumentation is mainly used for (highly reliable and routine) bulk analysis of materials, e.g., in industrial quality control laboratories and destructively in geological research. In the field of EDXRF instrumentation, next to the equipment suitable for bulk analysis, several important variants have evolved in the last couple of decades. EDXRF became commercially available in the early 1970s with the advent of high-resolution solid-state detectors (see Chap. 2). In principle, the high-geometrical efficiency of the semiconductor detector (usually silicon drifted detector) in EDXRF instruments, permits a great variety of excitation conditions. The final analytical capabilities and in particular the detection limits that can be attained by the instrument strongly depend on the sophistication of the detector electronics.

Types of X-ray sources used in portable/desktop XRF:

(a) *Sealed X-ray tubes.* Most commercially available X-ray spectrometers utilize a sealed X-ray tube as an excitation source, and these tubes typically employ a heated tungsten filament to induce the emission of thermionic electrons under high vacuum. After high-voltage acceleration, the electrons are directed toward a layer of high-purity metal (e.g., Ag, Cr, Rh, W, Mo, Rh, Pd, etc.) that serves as the anode. In the metal layer, the bremsstrahlung continuum is produced upon which the characteristic lines of the anode material are superimposed. The shape of the emission spectrum can be modified by changing the electron acceleration voltage. The broad band radiation is well suited for the excitation of the characteristic lines of a wide range of atomic numbers. The higher the atomic number of the anode material, the more intense the beam of radiation produced in the tube, and the more effective is the acquisition of lower atomic numbers.

(b) *Radioactive sources* are the most commonly employed in PXRF systems. Generally, these sources are very compact compared to X-ray tubes; a-sources are suited for the analysis of elements of low-atomic number . Frequently used sources are the ^{244}Cm, with a half-life ($t_{1/2}$) of 17.8 years that emits 5.76 and 5.81 MeV a-particles, and ^{210}Po, having a half-life of 138 days and emitting 5.3 MeV or the γ-rays emitting isotopes of ^{55}Fe ($t_{1/2} = 2.7$ years), ^{241}Am ($t_{1/2} = 433$ years), and ^{57}Co ($t_{1/2} = 0.74$ years).

With regard to detection systems in XRF technology in general, several detectors have been constructed, aiming at higher energy resolution and high count rate efficiency. A review of PXRF, including gamma- and X-ray detectors, by Knoll (2000) and Potts et al. (2001) presented the full chronological history of all the types of detectors since 1950 that have been used for these spectroscopic regions (see also Potts and West 2008). The authors included the latest trends in portable, thermoelectrically cooled and silicon drift X-ray detectors and the more recent high-resolution cryogenic X-ray detectors.

Of interest is the work of Langhoff et al. (1999) that promotes the combination of a high brilliance, a low powered X-ray tube, a capillary optical system, and a non-cryogenic X-ray detector for applications in the analysis of works of art. Useful frequently updated information on PXRF can be found in the biannual Newsletter of the International Atomic Energy Agency's (IAEA) Laboratories in Seibersdorf, Austria.

Here we present a brief outline of the available portable XRF devices for some archaeological applications, ,refer to the detected trace elements, discuss the comparison with other desktop and analytical methods, introduce some prominent applications with various cultural materials, and weigh the potentials, limitations, and future growth and prospects, with emphasis on Aegean and World obsidian provenance.

Applications in Art and Archaeology

Based on the new advancements on both the nuclear technology and the commercially available portable systems, the collection of quantitative data from archaeological samples is becoming one of the most trusted and widespread methods in archaeology.

The non-destructive capabilities of XRF are indeed particularly suited to research in art and archaeology, where the sample is unique or its integrity has significant technical or esthetic value. This is perhaps best exemplified in the examination of works of art, where the forensic aspects of the measurement may provide historical insight (Longoni et al. 1998; Spoto et al. 2000).

For example, if one assesses that the surface of archaeological and historical materials has deteriorated and differs in composition from the bulk, the quantitative measurements may require surface abrasion or even sampling, depending on the material; on the other hand, when a measurement has to be strictly non-invasive, one cannot expect reliable quantitative data. However, for each experimental situation, the choice of one approach over the other depends on the type and intrinsic value of the object, the aim of the investigation, the instruments available, and, last but not least, the scientist's or conservator's personal assessment of the acceptable damage.

Nonetheless, PXRF analyzers have shown during the last 30 years that they are ideal tools to aid in a variety of applications in cultural heritage, archaeology/geoarchaeology, and in archaeometry research in general (Rotondi and Urbani 1972; Hall et al. 1973; Cesareo et al. 1996, 1999; Longoni et al. 1998; Langhoff et al. 1999), including:

1. *On-site material characterization* with hand-held systems. Within the framework of a large scale geoarchaeological project in the area of eastern Crete, Greece, sediments from five drilling boreholes were analyzed using a PXRF; based on the resulted quantitative data, distinguishing of the fluvial or the marine character of the successively deposited sediments in all studied cores became feasible (Zacharias et al. 2009). In general, direct and in situ material characterization is a triggering task, and conclusive answers may come to light only with the use of a combination of calibrated portable sets that perform within an interdisciplinary environment of material scientists and geo-archaeologists.
2. *Museum analysis* using both hand-held and laptop systems. The PXRF can identify components of pigments in paintings and glazes, metal alloy content, provide the characterization of objects such as jewelry, silverware, and weaponry; thereby assisting conservators in the preservation and restoration of artifacts, as well as aiding in constructing databases from analytical data available for the scientific community.
3. *Provenance studies* based on the identified elements and their concentrations and by comparing sources and artifacts.
4. In *Conservation science* aimed at matching pigments in a rapid and accurate manner, helping to identify how objects have been preserved in the past, and

Fig. 6.1 Spectrace 9000 TN PXRF measuring mural paintings of modern artist Spyros Papalou-kas from the Amfissa Cathedral, near Delphi, Greece

provide the necessary knowledge in order to better conserve them for the future by examining elemental compositional data (Fig. 6.1).

5. In *art fraud or authentication testing*, the use of PXRF analysis performed on a variety of artifacts based on elemental data is extremely useful (Appoloni et al. 2007; Guerra 2008). Authenticating pieces helps prevent fraud, and can ensure that a returned artifact is the same one that a museum loaned or that donated pieces are identical in composition.

6. XRF has been used in combination with Raman spectroscopy in the identification of mineral pigments in modern art (Vandenabeele et al. 2000a; Liritzis and Polychroniadou 2007) and binding media and varnishes used in medieval paint-ings and manuscripts (Vandenabeele et al. 2000b).

7. *Ceramics.* Similarly, EDXRF was used for the examination of decorated shards of Neolithic pottery from Northern Greece (Papadopoulou et al. 2007). Using elemental concentrations for the painted and the non-decorated surface areas, the study revealed the main characteristics of each decoration, indicating the use of different techniques and exploitation of various clay deposits.

8. *Coins.* A comparison of PIXE and XRF for the elemental analysis of Japanese coins has been reported (Haruyama et al. 1999). It was found that selective

filtering was necessary to achieve successful PIXE analysis, and that XRF was the preferred approach. The relative concentration of low-level elements was found to be associated with their place of manufacture. The relationship between elemental composition of coins and their historical or geographical origin was the subject of a number of studies. These included the examination of Japanese medieval coins, 280 ancient Dacian coins from Romania (Cojocaru et al. 2000), Hungarian coins from the fifteenth century (Sandor et al. 2000), Ayyubid and Mamluk dirhams (AlKofahi and AlTarawneh 2000), and silver coins from the time of Alexander the Great (KallithrakasKontos et al. 2000). In the last case, the effect of the silver corrosion layer on the surface of the coins on the analytical results for 12 elements was studied before and after removal of the corrosion product.

9. *Manufactured glass.* Analysis of two sets of seventh to fourth century BC glass bead collections from Greece using a benchtop EDXRF set resulted in the characterization of the collection (Zacharias et al. 2008); in a subsequent step, the same collection was re-examined with the addition of one more calibrated EDXRF set and the introduction of the ion beam of PIXE, PIGE (particle induced gamma-ray emission), and RBS (Rutherford back scattering spectrometry) (Sokaras et al. 2009) provided good consistency and overall agreement between XRF and PIGER/PIGE for the detected major and minor elements. Application of radioisotope-excited XRF on eighth century Polish, Brandenburg, and Saxon glassware (Kunicki-Goldfinger et al. 2000), as well as EDXRF of Celtic glasses (Wobrauschek et al. 2000) has been reported. Furthermore, in porcelain analyzed by Wu et al. (2000), different compositional patterns were found for samples from various Chinese dynasties and for different usage. It was claimed that the method employed produced a highly efficient method of classification. In a similar study by Leung et al. (2000a), the use of principal components analysis of EDXRF data derived from 41 pieces of Dehua porcelain indicated that most of the samples were distributed in three areas, corresponding to the source of production.

PXRF Instrumentation

Following the increased demand for non-destructive XRF analysis, there are a number of options, depending on intended use. Within the family of solid state detectors, one has two main groups: (a) cryogenically (liquid N_2) cooled detectors and (b) thermoelectrically (Peltier effect) cooled detectors. Group (a) includes Si(Li) and HPGe detectors that provide good energy resolution (about 150 eV at 5.9 keV) and thick depletion layers, which means intrinsic efficiencies close to 100% for all the energies of practical interest (Knoll 2000) and large surfaces, that in turn mean high-detection efficiency. Typical figures for planar HPGe's are 13 mm and 200 mm^2 for thickness and surface, respectively. However, the need for liquid N_2 obviously reduces their portability and autonomy, and furthermore, due to large dimensions, the detector is positioned relatively far from the measurement point.

On the other hand, the detectors of group (b) do not need liquid N_2 and have small dimensions, which allow the detector to be placed very close to the measurement point; nevertheless, they have, with the exception of HgI_2 detectors, thin depletion layers – about 300 mm, corresponding to about 5% intrinsic efficiency for the Sn K-line – and small surfaces (less than 10 mm^2) that create poor detection efficiencies; the group includes HgI_2 detectors, with good energy resolution (about 200 eV at 5.9 keV) and intrinsic efficiency; SiPIN detectors, with poor energy resolution (about 250 eV at 5.9 keV) and detection efficiency; and Si drift (SD) detectors, with good energy resolution (about 160 eV at 5.9 keV, even for high count-rates) and poor detection efficiency (see Cesareo et al. 1992, 2008).

A brief review of some of these portable XRF equipments operating in the energy or wavelength mode is given below:

The EDXRF field portable analyzer Spectrace 9000 TN is operated with a mercuric iodide (HgI_2) detector, which has a spectral resolution of about 260 eV FWHM at 5.9 keV, and three excitation sources of radioisotopes within the probe unit – Americium $^{Am-}$241 (26.4 keV K-line and 59.6 keV L-line) measuring Ag, Cd, Sn, Ba, Sb; Cadmium Cd-109 (22.1 K-line, 87.9 keV K- and L-line) measuring Cr, Mn, Fe, Co, Ni, Cu, Zn, As, Se, Sr, Zr, Mo, Hg, Pb, Rb, Th, U; and Iron Fe-55 (5.9 keV K-line) measuring K, Ca, Ti, Cr. The wide range of excitation X-ray energies theoretically can measure all chemical elements from $Z = 16$ (sulfur) to $Z = 92$ (uranium); however, a variety of limiting factors produce a lower threshold of detection and in practice 26 simultaneous elements are measured.

The performance of the portable Spectrace 9000 TN EDXRF instrumentation in the laboratory has been reported (Potts et al. 1995; Liritzis 2007), while the entire concept of most investigations may apply to any available portable XRF instrument that requires an irradiated area rather than a point focus. Reference samples included rhyolites, metals, clays, soils, and low radioactivity reference samples. Several studies showed the capability of the instrument to determine major and minor elements (K, Ca, Fe, Ti, Mn) and selected trace elements (Rb, Sr, Zr, Mo, Ba, Pb) in typical silicate (rhyolitic) rocks. Other trace elements are not measured because their lower counting sensitivities mean that the concentrations were near to or below detection limits. The disadvantage of such PXRF with radioisotope sources is the need for replacement of the excitation sources that provide the primary X-rays every few years due to limited half-lives, which may become a serious problem when the production company no longer provides such sources.

Bruker company provides the Artax benchtop system that performs a simultaneous multi-element analysis in the range of Na(11) to U(92), by operating in two filtered modes using tungsten and molybdenum. In the latter, the X-ray tube generates 2–5 times larger peak areas for K-line elements above 20 keV for a better detection of light elements. The set is a micro (μ) XRF system since it provides a spatial resolution of down to 70 μm with the use of a CCD camera for a magnified digital image of the sample region under investigation, a white LED for illuminating the sample and to optimize the image quality and a laser diode to control the exact position of the beam on the sample and the exact distance between object and spectrometer.

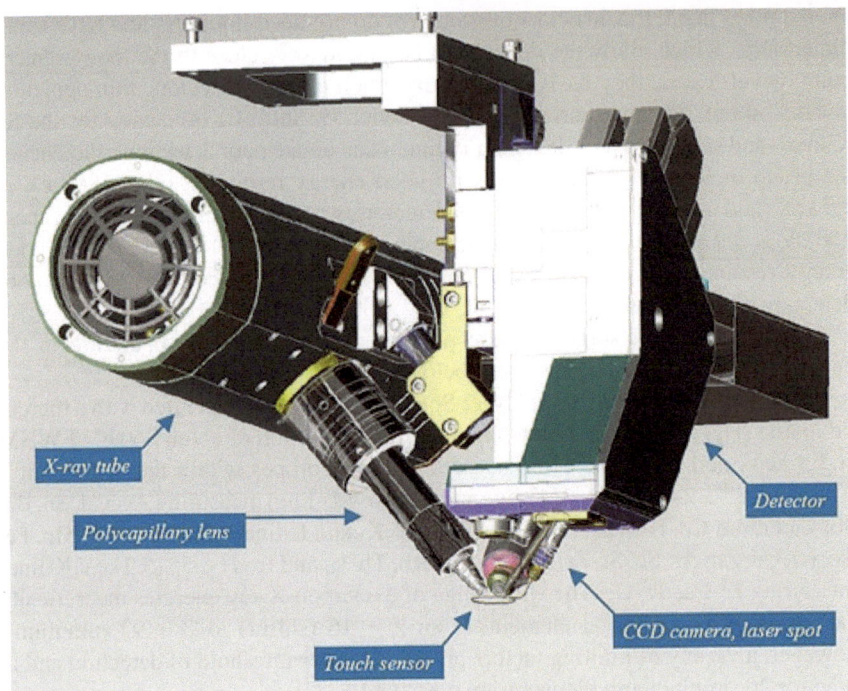

Fig. 6.2 Bruker handheld portable μ-XRF device Artax – second generation (http://www.bruker-axs.de/artax.html)

The *handheld Tracer* (*turbo-SD*) *option from Bruker* incorporates silicon drift detector technology, thus additionally providing the concentrations of Mg, Al, and Si without vacuum or helium attachments (Fig. 6.2).

Thermo scientific provides a series of analyzers either for metal alloys or for multi-elemental archaeological applications: The extremely light (1.3 kg) battery operated NITON XL3t with a X-ray tube (Au anode at 50 kV) also provides the concentration of ca. 25 elements from sulfur to uranium, plus Mg, Al, Si, and P when connected via helium flush.

NASA Goddard Space Flight Center PXRF unit operates at variable voltages of up to 60 kV and variable currents of up to 100 μA, obtaining accurate results quickly. This unit uses a metal-ceramic X-ray tube and a cadmium tellurium (CdTe) detector. Calibration can be automated, and the unit can be designed to include data accumulation, processing, storage, and transmission systems.

Oxford Instruments (*X-MET5000*), *Innov-X, and Spectro xSort,* all specialize in mining and metals operation solutions by providing handheld and benchtop sets for an effective measurement of the above-mentioned range of elements.

In contrast to the given information for some of the above systems that can reliably detect concentrations of light elements ($Z \leq 13$), this still has to be exploited in more archaeological applications and to be supported with published reference data.

In many XRF research laboratories, portable XRF spectrometers are built from commercially available components and in-house patents; such a μ-XRF set has been developed at N.C.S.R. Demokritos (Athens) and is described in Karydas (2007).

A portable XRF spectrometer with polycapillary optics and vacuum chamber has also been constructed at the IAEAs XRF laboratory in cooperation with the Atominstitute Technical University, Vienna, operated up to 50 keV and a focal size of 100 μm and a Silicon drift detector, for elements from Na upward (Wegrzynek, 2005).

The quantification procedure of intensity to concentration is based on the so-called fundamental parameter (FP) approach in XRF analysis (Ebel 1999; Kitov 2000). In general, the quantification of the spectral peaks to concentration is not simple, and care should be exercised. In fact, the FP correction technique is usually employed to calculate the true element intensity from the measured data. The efficiency model incorporates the bulk detector efficiency and the contributions from metal contact layers, dead layer and the beryllium entrance window. But the choice of which *calibration and matrix correction expression* to use has always been the subject of discussion. Thus, the assertion that the fundamental algorithm really was the most sufficient and the one from which all other theoretically valid expressions could be derived was the debatable issue (Willis and Lachance 2000, 2002). These authors questioned the original conclusions and strongly defended the use of the expression whose correctness had been called into question. Needless to say, such discussions will continue, while practitioners continue to use successfully the expression and software that they have validated for their own application.

Finally, the important topic of matrix correction using the FP approach continues to generate interest both in desktop and portable XRF applications. The detailed mathematics and calibration features of the FP method and some of the practical approaches that can be adopted for the solution of the FP correction equations were reviewed in detail by Kitov (2000). It is normally the case for FP correction algorithms that the excitation spectrum is either measured or calculated and calibrated.

Overall, the features of different portable XRF spectrometers are considered with special regard to the choice of X-ray tubes and detectors; this choice affects both portability and analytical performances. Moreover it has been shown that good detection limits are essential to investigate archaeological and historical materials since in most cases trace elements provide more information than the major ones.

Detected Elements and Clustering Techniques

The chemical elements usually measured by PXRF are K, Ca, Ti, Mn, Fe, Zn, As, Rb, Sr, Zr, Ba, Hg, and Pb. For cluster processing, only concentrations well above the detection limit and thus low error are used.

Two of the most popular multivariate statistical techniques used in archaeometry are cluster analysis (CA) and principal component analysis (PCA) (Baxter 1994). The application of CA should be applied to more or less homogeneous samples regarding origin and time.

Hierarchical CA were among the earlier multivariate approaches to be systematically deployed by archaeologists, and Pollard (1986) gives a brief discussion of applications to the provenance of pottery using chemical compositions. CA inheres a wide range of techniques (Baxter 1994; Mantzourani and Liritzis 2006).

At any rate, a common problem with all methods is the assessment of cluster validity, for example, to decide how many "real" groups are in the data; not an easy issue, as no generally applicable solution to determining the appropriate number of clusters exists. Complementary methods that contribute to this problem include PCA, correspondence and discriminant analysis, and model-based multivariate mixture of normals (Papageorgiou and Liritzis 2007).

The CA results are commonly presented as dendrograms showing the order and level of clustering, as well as the distance between individual samples. Statgraphics 5 Plus (Manugistics, Inc. 2000) and Splus packages are used. However, it is shown that simpler scatter plots corroborated by CA are sufficient to differentiate at least several obsidian sources from the World, with a high confidence (see below).

Comparison with Other Benchtop Methods: Elemental Differences

When comparing PXRF with other methods in the determination of similar major and trace elements, several issues must be taken into account:

1. About ±10–15% and for some elements 20–30% differences already are reported between the different desktop methods used on same samples. This can be explained in different ways. First, in the comparison of measurements obtained by two methods on aliquots of a single sample, one has to consider both the precision but also the accuracy of each one. About precision, one has a certain number of repeats on different samples. About accuracy, it may depend on the standards used in each method (i.e., U.S.G.S. standards are powdered, not satisfactory for PIXE or EMP-WDS).
2. The mass/volume of sample analyzed. Typically 100 mg by ICP; "punctual measurement" by EMP-WDS ("defocused" beam, to a diameter of about 5 μm, to avoid volatile elements losses); but a larger volume by SEM-EDS (scanned surface area of about 1 mm^2); even more by PIXE (scanning surface of about 1 mm square but a penetration depth within the sample by the proton beam of about 50 μm).
3. The glass composition homogeneity.
4. The nature, size, relative, and absolute frequencies of mineral phases.

5. Certain elements (a) are more precisely analyzed by one method than by another, and (b) in some cases for a given element one can be near detection limits (hence derive a larger error in content determination) for one method but not by another one, etc. The latter is an obvious issue for U and Th in PXRF.

6. For different samples, internal variability of a given source composition occurs. Moreover, internal variability within one single source is of significance too. This, for example, especially applies in the case of pottery analysis. Here, clay's inhomogeneity, caused either by the presence of non-normally distributed inclusions (Buxeda i Carrigos et al. 2003) or simply incomplete refinement processes, results in distribution issues. In pottery studies, it is also the high- or over-firing that can occur on a sub-set of artifacts produced at the same kiln that is responsible for alteration effects in the rare earth element concentrations (Schwedt et al. 2004) and therefore severely affects the compositional groups resulting from the chemical analysis (Tite 2008) and, naturally, any comparison between PXRF with other analytical methods.

Another cause of variability is the grain size of the ground powder as well (see below for obsidian); but if inter-source differences are large, there may be no problem in source assignment.

Thus for sourcing, absolute element contents are not necessary, for example, see the works by Acquafredda et al. (1999, and Chap. 2) on XRF, where they use only peak intensities.

A past review of the relative merits of PIXE, XRF, and ICP-MS in the analysis of archaeological artifacts (Pillay 2001) and of EDXRF and X-ray microanalysis (Linke and Schreiner 2000) may also be of interest in this context.

Worldwide Examples

Pigments

The identification of pigments and inorganic materials used in works of art is fundamental to further the understanding of an object's history or an artist's technique, and may provide evidence for dating or attribution of artifacts. Characterization of the artist's original materials as well as materials applied later (by artist, conservator or forger) is useful for providing criteria for conservation decisions.

The onsite XRF technique is, however, subject to some intrinsic limitations. The so-called matrix emission intensity of each element is a function not only of its concentration but also of the overall composition of the larger area under investigation. Moreover, the technique is capable of detecting only the elements and not the compounds to which they belong, and an unambiguous identification is not possible for numerous copper, lead, cobalt, and chromium containing pigments. These limitations can be particularly important when XRF analysis is applied to the study of the modern artists' palette that may comprise natural and synthetic pigments as well as complex mixtures, either mixed by the artist or used as readymade tube paint-formulations. Indeed, from the nineteenth century onward, paintings have become more complex in terms of the number and mixtures of materials used for a single work. Newly invented pigments became available, which were sold already mixed in tubes allowing artists to free themselves from traditional studio practice and to render their subject matter more vividly.

Desnica et al. (2008) investigated the pigments from the painted wooden inventory of the pilgrimage church of Saint Mary of Jerusalem in Trski Vrh – one of the most beautiful late-baroque sacral ensembles in Croatia. It consists mainly of two painted and gilded layers (the original one from the eighteenth century and a later one from 1903), partly overpainted during periodic conservation treatments in the past. The approach was to carry out extensive preliminary in situ pigment investigations using PXRF, and the problems not resolved by this method on site were further analyzed ex situ using μ-PIXE (particle-induced X-ray emission) as well as μ-Raman spectroscopy. Therefore, the XRF results acted as a valuable guideline for subsequent targeted sampling actions, thus minimizing the sampling damage.

In Agnoli et al. (2007), the analysis of samples of Roman Age mural paintings was carried out by using energy dispersive X-ray spectroscopy, XRF, and PIXE, resulting in the identification of the pigments used; in Capitan-Vallvey et al. (1994), the technique was used to identify the various pigments that appear in the decorations of the "Corral del Carbon" in Granada a fourteenth century restored monument.

Another comparison study on pigment identification was that of Perez-Arantegui et al. (2008) where PXRF performance was compared with laser ablation-inductively coupled plasma mass spectrometry (LA-ICPMS) for the characterization of cobalt blue pigments used in the decoration of Valencian ceramics. Qualitative data on the elemental composition of the blue pigments obtained using both techniques show a

Fig. 6.3 Anthivolo, preparatory drawing of pinched paper used by modern artist Spyros Papaloukas to paint Amfissa Cathedral

good agreement. Moreover, the results clearly illustrate that potters utilized different kinds of cobalt pigments in different historical periods.

Another interesting application of PXRF has been made on the wall paintings (hagiography) and preparation drawings (*anthivolo*) of the Amfissa Cathedral (Central Greece) by early twentieth century artist Spyros Papaloukas (Fig. 6.3).

Sampling locations ranged from the ceiling to ground level and on various decorative motifs and compositions. Only minute samples (at millimeter scale) were permitted. Any perceptible invasive sampling was unacceptable. The non-destructive EDXRF measurements were performed on hagiographic representations in the pulpit, the women's gallery, and the left aisle (Liritzis and Polychroniadou 2007). Sampling was restricted in view of the value and the condition of the wall paintings. Pigments identified from elemental concentration included red and yellow ochres, iron oxides, and lead and titanium white. Stains of similar pigments were also found in the *anthivolon*. In particular, the paper support, red and black stains, were analyzed by EDXRF, FTIR, and Raman. The black stain contains mostly iron probably an unknown red paint poured accidentally on the tamping (tampon) layer of the prickled paper which contained sienna, chrome yellow, white of titanium, and others. The red stain contains much more iron identified as iron oxide (sienna). The rest of the paper contains faint yellow, traces of sienna, chrome yellow, and lead oxide (orthorhombic massicot). See Tables 6.1 and 6.2.

Non-destructive analysis for pigment identification of colors on two pieces of anthivolo using Raman and EDXRF methods indicates that EDXRF yields only chemical element concentrations, while Raman spectroscopy offers information about the structure (Liritzis and Polychroniadou 2007).

Sample location, color, method used, and pigment identification by different methods on minute samples from various points of the wall painting is shown in Table 6.2 (high barium contents as $BaCO_3$ white are present rather due to weight increase and thus higher cost). Portable EDXRF analyses focus in a non-destructive manner on a larger area on the surface mural painting. Chemical elements are reported that reflect known compounds, in contrast to the tiny samples detached and analyzed by Raman and FTIR. The compounds here are recognized from a data bank spectral library. Manganese and iron oxides (ochres) and minion were used in antiquity.

Table 6.1 The results of combined EDXRF and Raman analysis for the identification of colors on two pieces of anthivolo, drawing paper

Anthivolo sample reference	Color	Pigment	Method
AMFA-1 (26 × 56.5–59 cm)	Yellow	Lead chromate sulfate	Raman, EDXRF
	Black	Carbon	Raman
	Black stain	Contains iron (sienna)	EDXRF
	Red stain	Iron oxide (sienna)	Raman
		Iron oxide + titanium dioxide white (TiO_2)	EDXRF

Table 6.2 Sample data, color description, pigments identified and the techniques applied

Sample reference	Description	Color	Pigment	Method
AMF2	ΓΔ3, women gallery, right aisle third apse, yellow flaked pigment	Yellow	In four successive painted layers: chrome yellow + yellow ochre + ultramarine + red ochre (yellow ochre=kaolinite by FTIR) High content: Pb and Fe (mainly yellow from similar pigments as above)	Raman, FTIR EDXRF
AMF3	ΓΔ2,women gallery, right aisle, second apse, red flaked pigment	Red	Iron oxides + Ca + Sr. Red ochre	EDXRF
AMF4	Pulpit, red decoration	Red	High Fe, Ca, Sr, Ba. Red ochre	EDXRF
AMF5	Left aisle, internal side of column	Blue	High Fe, Ba, $CaCO_3$, K, Prusian blue	EDXRF, FTIR, IRPAS
AMF12	ΓΔ3 Women gallery, right aisle, third apse)	Blue	High Fe and $CaCO_3$. Red ochre	EDXRF
AMF14	Pulpit, part of surface, flaky pigment on paper with lettering	Yellow	Yellow of Pb-Sn + PbO, K, F (a surprising result which cannot at present be justified)	EDXRF
AMF17	ΓΔ3, women gallery right aisle, third apse red from the vault	Red	Red lacquer + Red ochre Iron oxides + MnO_2 + Clay. Red ochre	Raman EDXRF
AMF18-AMF26	Local shop	Various colors (modern pigments)	Gypsum, minium, oxides of manganese and iron, high barium	EDXRF

PXRF can assist in archaeometric issues when studying decorated surfaces of archaeological pottery fragments (Papadopoulou et al. 2006; Romano et al. 2006) and in Cathedrals (Ferretti et al. 1991).

However, limitations due to X-ray attenuation by absorption make the technique ideal for smooth surface analysis. Care must be taken to avoid any secondary products from weathering/erosion or corrosion effects; for very thin layers, such as the painted layers, the obtained data always provide semi-quantitative results, while in cases of multi-layered compositions, the recorded spectra are composite from all the layers in the vessel.

Metals

Portable and micro-XRF operations are critical in ancient metal studies for the identification, characterization, mapping, and thickness determination of both alloys and, more important, the corrosion products; these products, usually distributed within small areas of the artifact surfaces, are strongly associated with the archaeological environment as well as some indoor conditions (Ferretti and Moioli 1998; Ferretti et al. 1997).

Thus, in archaeometallurgical studies, chemical analysis should work in line with metallography, and the conservation knowledge and background of the artifact. In the event that any corrosion or patina product is removed, the X-rays should be directed to the pure metal phase. When only the surface coatings are analyzed, the results are suspect.

In fact in quantitative EDXRF analysis of ancient metallic objects, two main difficulties emerge: (1) determining the correction factors for the irregular shape or relief effects, and (2) measuring the true composition of the bulk metal under the surface patina. In the case of coins, taken as a typical example, point (1) could, in principle, be by-passed by casting pure metal copies of specimens and comparing XRF intensities with the ones from regularly shaped standards. The interest in examining coins, however, mostly depends on the possibility of the analysis method to be applied to several pieces, and XRF analysis should prove to be impractical in this case. Gold alloy objects do not normally present a patina on the surface, so by choosing proper geometric conditions during irradiation and by resorting to XRF line intensity ratios, it is possible to eliminate the problem of evaluating geometric factors. Chemical analysis of metals and metal alloy collections provides information on the manufacturing process, the provenance of raw materials, the geographical distribution of the ancient metallurgical technology, etc.

PXRF was used for the in situ study of gold and silver jewels (seventh to first century BC) from the Benaki Museum of Athens (Karydas et al. 2004). For the gold objects, the use of two distinctive sources was revealed - a, native and of high purity gold alloy and for the silver jewels analysed the copper content was evaluated as a technological, parameter of the materials while the presence of some minor elements like Pb, Bi, and Au, was attributed to an argentiferous galena used for the silver production and thus confirmed the authenticity of the jewels. Within the frame of the same study, parallel use of PIXE and PXRF spectrometers was made for three red gemstones on jewels (fourth century BC to first century AD) exhibited at the same museum (Pappalardo et al. 2005). The analysis showed that one out of the three red stones analyzed was a pyrope-type garnet and the other two were the almandine type. The use of PXRF enabled the determination of trace elements, such as Cr and Y, which turned out to be the decisive factor in the classification of the red garnets into different types; India and Sri-Lanka were then proposed to be the geographical provenance for the analyzed red garnets.

In authenticity testing of gold jewelry, the analysts should consider a number of difficulties that arise mainly from the non-destructive examination to be applied. Initial observations using microscopy techniques, such as optical microscopy, scanning electron microscopy and X-ray radiography, are used to determine manufacture techniques. Then, the in situ elemental analysis using portable μ-XRF devices are inevitably used to provide the concentrations of the major elements of gold alloys including gold, silver and copper. Moreover, technological changes can be attributed to economic and seasonal differentiations while the ratios of some characteristic elements of gold (Pd, Sn, Sb, Pt, and so on) are straightforwardly correlated to the geological origin.

In Vivo PXRF Analysis

Another interesting application of PXRF can be found in Rebocho et al. (2006). In this work, the post-mortem lead concentration in human bones of the Middle Age was measured by means of a PXRF system based on [109]Cd radioactive source consisting of a Ge hyper-pure detector. This system, conceived for in vivo Pb analysis in bone, is portable, non-destructive, and based on lead K lines detection in contrast to the common ones based on the L spectra, in the usual X-ray fluorescence techniques. The drawback of this technique is that only elements of high atomic number are detected with enough efficiency and resolution, while the other technique allows simultaneous detection of most of the sample elements. Furthermore, this work has highlighted that diagenetic alteration in bone depends mostly on the physical and chemical properties of the burial place and the structure of the bone.

Combined Portable XRF with XRD and μRAMAN

Since the beginning of this century, instrumentation allowing both chemical and compound analysis appeared in scientific literature. In Uda et al. (2005), a portable X-ray diffractometer equipped with an X-ray fluorescence spectrometer was set up so as to get a diffraction pattern and a fluorescence spectrum simultaneously in air from the same small area on a specimen. Diffraction experiments were performed in two modes, i.e., an angle rotation mode and an energy dispersive mode. In the latter, a diffraction pattern and a fluorescence spectrum were simultaneously recorded in a short time, 100 s or less, on one display. The diffractometer was tested in the field to confirm its performance. Targets chosen for this purpose were a bronze mirror from the Eastern Han Dynasty (25–220), and a stupa and its pedestal which are part of the painted statue of "Tamonten holding a stupa" from the Heian Period (794–1192 AD), enshrined in the Engyouji temple founded in 996. The performance of the diffractometer equipped with XRF should be improved in the near future by installing a two dimensional scanning stage and by introducing computer software for quantitative analysis.

Another promising combined facility is that of a mobile micro-analytical μRaman and XRF set up that was built within the PRAXIS European project (Andrikopoulos et al. 2006); the instrument permits structural characterization of the pigments incorporated in a painting under study by evaluation of the Raman spectra, together with elemental analysis of the same materials provided by XRF spectra. The validation of the instrument's in situ and non-destructive capabilities was performed after its in vitro application on an experimental icon (painted with traditional Byzantine techniques). The data acquired by the two techniques from the same areas on the painting offer complementary results, which enable the identification of almost all pigments even in the case of over-painted art objects.

Ceramics

Ceramic and similar material, such as bricks and clay deposits, were usually exploited analytically with the use of standardized multi-elemental techniques like NAA, ICP, and laboratory XRF infrastructures. But nowadays due to the improvement in the elemental range measured by PXRF, the use of portable applications can be increasingly seen in the literature. PXRF has been applied extensively to archaeological materials such as ceramics, clays, soils, focused on clay provenance, clay fabric similarities, and trade exchange issues processing the data by clustering techniques, for example, Mantzourani and Liritzis (2006), Papadopoulou et al. (2006), Papageorgiou and Liritzis (2007), Liritzis et al. (2002, 2007), Liritzis (2005), Pappalardo et al. (2003).

Other Applications

1. Uda et al. (2000, see also Uda et al. 2005) analyzed some pigments on the Funerary Stele of Amenemhat (ca. 2000 BC) exhibited in the Egyptian Museum, Cairo, and on the walls of a rock-cut tomb in Thebes, Egypt. Measurements were made with a home-made XRD instrument and a commercial PXRF under touch-free conditions. Hunite (a white Ca–Mg carbonate pigment) and an As-bearing yellow pigment were detected.

2. In another archaeological application, a portable EDXRF instrument incorporating a calcium or lead-anode X-ray tube was used by Cesareo et al. (2000) to determine Cl and S in frescoes and stone monuments. Detection limits were reported to be 0.04% m/m for Cl and 0.03% m/m for S.

Provenance Studies of Obsidians by PXRF

Provenance studies of the raw materials used by prehistoric lithic industries are of key importance in research on ancient man. They provide basic information on the extension of the territory exploited by small groups of hunter-gatherers during the Paleolithic age. With respect to the Neolithic period ((seventh centuryto fourth millennia BC), provenance studies contribute to the knowledge of long-distance circulation and exchanges of raw materials, using *chaines operatoires* of lithic artifacts. Indeed, reconstructing mobility strategies is a major goal of researchers interested in prehistoric hunter-gatherers, and the use of geochemical source characterization of obsidian found at sites in a region offers a way to reconstruct the procurement range, or distance traveled to obtain resources of prehistoric groups (Roth 2000; Leslie et al. 2007; Craig et al. 2007; Shackley 2005; Tykot 2001).

Obsidian, due to its often remarkable knapping properties and esthetic qualities, was frequently used by prehistoric people. It is also one of the preferred materials in provenance studies. This is because of its mode of formation in volcanic events and its glassy matrix; in addition, the physico-chemical properties of an obsidian are most often similar even at a micro scale.

Early obsidian provenance studies were based on bulk physical properties, such as, color, density, refractive index, etc., as well as on petrography. Although useful for sample description, these observations generally do not provide valuable criteria for provenance studies (Gopher 1983).

However, for archaeological obsidian artifacts, which generally have to be studied non-destructively, X-ray fluorescence has proved effective in analyzing large numbers of artifacts in a short time for characterization and provenance (Davis et al. 1998; Potts et al. 1995, 2001, and for provenancing British stone axes (Williams-Thorpe et al. 1999).

For example, recent analytical, dating, source, and trade studies within the western Mediterranean, central and Eastern Europe, the Aegean, and Anatolia and the Near East reviewed by Williams-Thorpe (1995) and De Fransesco et al. (2007), have increased the use of PXRF on obsidian characterization and analysis in the region. Results of these studies have shown that distributions are mainly separate in the four regions examined, and that obsidian was traded for distances up to 900 km in the prehistoric period.

As chemical sourcing is becoming an increasingly important component of archaeological investigation, PXRF leads in this field of research. As discussed in Chap. 2, Craig et al. (2007) used the technique with laboratory EDXRF on 68 obsidian artifacts from the Formative site of Jiskairumoko, in southern Peru. Both techniques arrived at substantively similar conclusions.

Although a powerful approach, a geochemical characterization does not in all cases allow one to discriminate obsidians from different volcanoes or from different lava-flows within a single volcanic field, so the choice and measurement of particular elements is vital.

The classification of Aegean obsidian sources and artifacts (see below) by well calibrated portable EDXRF and simple scatter plots, complemented by 3D plots and dendrograms, offers the advantages of portable XRF, for example, (a) swift counting time, (b) analysis of tools by a non-destructive measurement, (c) low cost, (d) versatile as a portable analyzer performing in situ, (e) groupings by simple elemental biplots, and (f) comparison with other methods.

In the following sections, sample mineralogy (size and distribution), but also the percentage covering over a 25 mm diameter analyzer window with solid flat artifacts, as well as, correction factors will be defined (Liritzis 2007), while determination of critical powder thickness in vial sample holder, for lighter and heavy elements, provides a test for correct concentration of certain elements.

Effect of Grain Size on Elemental Concentration

The effect of grain size of ground obsidian on elemental concentration has been examined. The grain sizes of <32, $32-71$, $71-90$, $90-125$, $125-150$, $150-212$, and $212-500$ μm were collected, through a mesh, from obsidian flakes of Melos island, Greece. The general trend is reduction of concentration as a function of increasing grain size. For the two sources (Adamas and Demenegaki) and the candidate location at Katsouli (may be a working location), in Melos, the obtained differences in the trends for some elements are the following: For K the differences between the extreme sizes are 15–20%, for Fe 25–30%, for Ti 18–30%, and for Ca for both sources are around 19%. Moreover, the elemental concentration for some obsidian powders was compared to data obtained by respective solid flat samples. Similar concentration values within ±5% were observed for K, Ca, Sr, Zr, Rb of Demenegaki and Adamas sources, while for Ti powder is around 25% lower than solid, and for Fe around 20% higher than solid.

Dependence of Atomic Number (Z) of Elements on Critical Depth and Detection Limits

It is well known that the sample thickness affects elemental precision, particularly at high energies (Davis et al. 1998, Chap. 3). Hence it is useful to have a measure of appropriate thickness for certain elements, particularly in case of having small amounts of powder samples or thin solid samples.

The data of critical depth Dc vs. atomic number (Z) of eight elements were obtained from Potts et al. (1997). They calculated Dc (in μm) for 99% of the secondary X-rays for the eight elements of three reference samples (basalt BCR-1, andesite AGV-1, and rhyolite RGM-1). These samples were measured by our Spectrace 9000 TN portable EDXRF too, for heavy elements K, Ca, Ti, Fe, Rb, Zr, Ba, and Ce which are mostly affected from sample thickness.

The aim was to construct a relationship between Dc, in which secondary X-rays reach within the sample, but for 25 elements detected, and their Z. For this, plots between Dc and $1/Z$, as well as, against $1/Z^2$ per each of eight elements used by Potts et al. (1997) were made. The latter plots fit the eight data points much better than simply vs. Z. Table curve 2D was employed and various fitted curves produced, chosen the one with best r^2 closest to one. Verification of the appropriate equation was made using well-known data through Mathcad 2000. The best equation is the $1/Z^2$ dependence of Dc given in (6.1).

$$1/Z^2 = a + b \ln x + c \ln^2 x + d \ln^3 x + e \ln^4 x + f \ln^5 x + g \ln^6 x, \qquad (6.1)$$

where $x = $ Dc in μm, and a–g numerical coefficients (see Table 6.3).

Figure 6.4 shows significant differences in Dc below 0.0012 ($>Z = 29$) for these three rock types, implying the need for having thick samples of certain size per rock type. For example for $Z = 38$ (Sr), the Dc between basalt and rhyolite differs by about 0.7 mm.

Following similar concept, a relevant curve was made for soil. The Dc values are taken from Operation Manual of Spectrace 9000 TN portable analyzer. Elements used are Na, Si, Ca, Fe, Rb, Nb, Rh, La, and Eu (Fig. 6.5). The derived relationship was:

$$1/Z^2 = (a + cx + ex^2 + gx^3)/(1 + bx + dx^2 + fx^3 + hx^4), \qquad (6.2)$$

where $x = $ Dc in μm.

Coefficients for (6.1) and (6.2) are given below in Table 6.3.

For soil and at the turn point of 0.001, the Dc increases significantly for measuring precisely elements with $Z > 32$. Therefore, for a particular rock type, construction of similar curves satisfies sample thickness and accurate concentration values.

Comparing Dc (99%) between basalt and soil, for say $Z = 40$ or zirconium ($1/Z^2 = 0.0006$), sample thicknesses are as follows: 0.946 mm for basalt, 1,784 mm for rhyolite, and 17.74 mm for the soil.

Application of this notion to actual concentration measurements has been made on reference radioactive sample BL4 (uranium ore, certified reference material CANMET, Canada) for K, Ca, Fe, Ti, Mn, Sr, Mo, Zr, Pb. Ba, U, Th as a function of sample thickness. Above 5 mm Basic elements above 5 mm are precisely measured. For heavy elements, e.g., Ba and Th, the Dc, and thus sample thickness, should be at least 13 mm for accurate results (Fig. 6.6).

Table 6.3 Calculated coefficients of (6.2)

	a	b	c	d	e	f	g	h
Basalt	0.147	−0.141	0.056	−0.012	0.001	-8.12×10^{-5}	1.983×10^{-6}	—
Andesite	0.126	−0.117	0.046	−0.009	0.001	-6.003×10^{-5}	1.422×10^{-6}	—
Rhyolite	0.095	−0.0867	0.033	−0.0067	0.00074	-4.274×10^{-5}	1.0056×10^{-6}	—
Soil	0.01	0.00012	−0.00042	0.0019	1.426×10^{-5}	5.254×10^{-5}	2.669×10^{-8}	3.427×10^{-6}

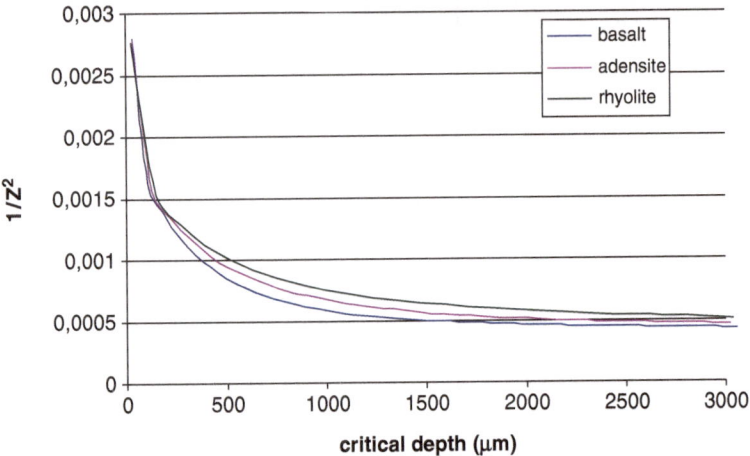

Fig. 6.4 Dependence of the atomic number (Z) vs. critical depth (Dc) for basalt, rhyolite, and andesite. This way, constructed family curves are made for particular element (Z) in a rock (basalt: *lower*, andesite: *middle*, rhyolite: *upper*)

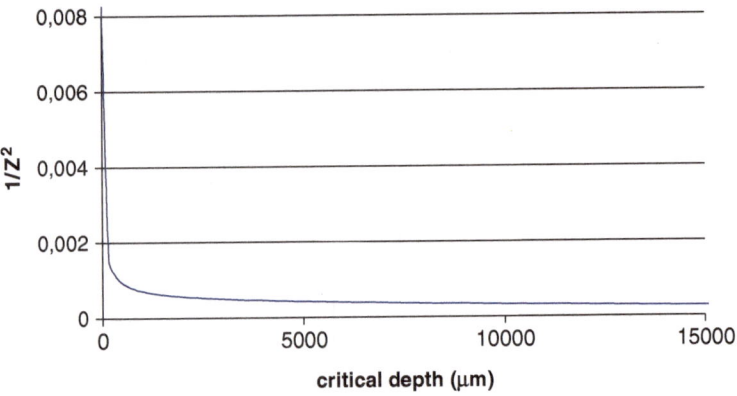

Fig. 6.5 Dependence of Z as a function of Dc in soil

Correction Factors for Measuring Small-Sized Obsidian Tools in Window-Type Apparatus

In window-type apertures of PXRF, the element concentrations derive from the total area, and the built-in software requires full counting geometry. In order to accommodate and measure accurately flat samples of size smaller than the window aperture, a procedure has to be devised (Bellot-Gurlet et al. 2000). For example, a procedure was devised relating the sample of variable sizes as percentage coverage of the window, of 25, 50, 75, and 100%, and the rest was covered with Perspex, the Spectrace 9000 TN portable EDXRF. Apart from the percentage coverage, the

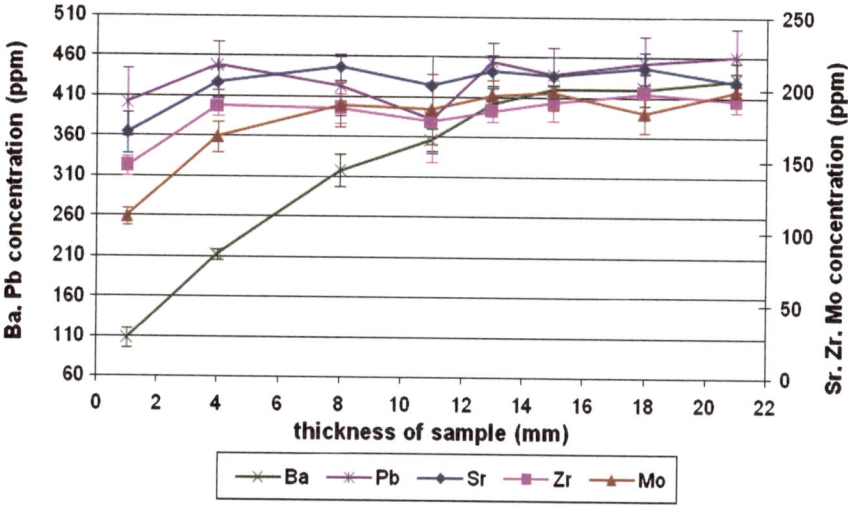

Fig. 6.6 Variation of K, Ca, Fe, Ti, Mn, Sr, Mo, Zr, Pb. Ba, U, Th, for standard sample BL4 as a function of sample thickness

position of sample is important, that is, if the cover is on the left, right, up or down, looking on the window, due to the irradiation and counting geometry. Obsidian from Adamas (Melos Island) was cut flat, and six measurements were carried out for each data point for 200 s in each reading, with a repeated interval of 2 years. The repeat measurements were very similar to initial ones, and the average was used. It was found that for the two positions, forward as we look at it (up) and toward the operator (down), there is a discrepancy with regard to the other two positions (left and right), which give similar results (Fig. 6.7). From these data, it was possible to produce correction factors for small-sized samples. These correction factors f (=total coverage/partial coverage) for left and right positions per element form a set of family curves.

Figure 6.8a–c shows three correction curves for K, Ca and Ti, and Fe, Sr, Zr, Rb, Ba.

Clustering Techniques of Aegean and World Obsidians

Any clustering technique is meaningful if the analysis is correct. PXRF may sometimes produce dubious results. This may happen in case of elemental biases, concentration values near threshold of detection, inappropriate calibration and lack of interlaboratory comparisons. When these parameters have been adequately controlled and secured, then data normalization or standardization is the choice and the use of some hierarchical cluster analysis techniques.

Figure 6.9 shows the clustering of prehistoric ceramic shards derived from several settlements from the Greek mainland and the Aegean, employing cluster

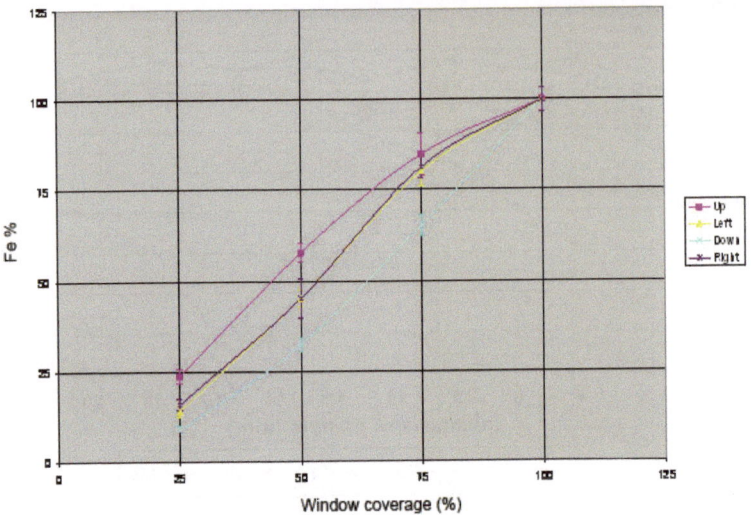

Fig. 6.7 Dependence of Fe concentration as a function of percentage coverage for four positions for the Spectrace 9000 TN

analysis with model-based multivariate mixture of normals (i.e., of data with normal distribution) (Papageorgiou and Liritzis 2007).

However, the groupings can also be resolved using either 3D plots (Liritzis et al. 2007), biplots of elemental ratios, or simple biplots. A case study of simple biplots applies to obsidians and their elements Sr and Ti – a simple but effective and fast result of PXRF (Liritzis 2007).

Figure 6.10 shows Sr vs. Ti (in ppm) of some Mediterranean sources and newly analyzed artifacts and Fig. 6.11 similar for some World obsidian sources acquired with our PXRF or by other desktop devices taken from the literature.

Discussion of Advantages and Limitations

From the above outline of portable and non-destructive XRF analyzers, a summary of the advantages and limitations can be outlined.

Limitations where particular attention is needed include: (1) PXRF cannot differentiate superimposed thin painting layers, (2) the presence of corrosion layers provides misleading results, (3) the thickness (infinite thickness) of sample is vital for some heavy elements, (4) flatness of sample. In fact, one of the well-known difficulties with in situ PXRF is that results are affected by surface roughness effects, more so than desktop EDXRF instruments. When analyzing whole samples, the effect of surface roughness usually introduces an additional air gap between sample and PXRF analyzer. In addressing this problem, Gauvin and Lifshin (2000) developed a Monte Carlo

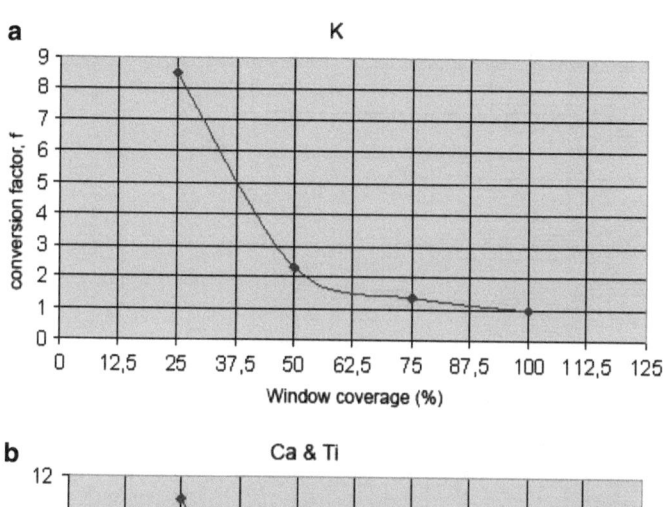

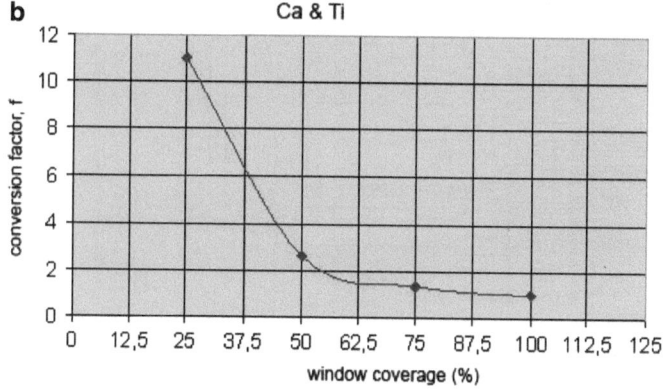

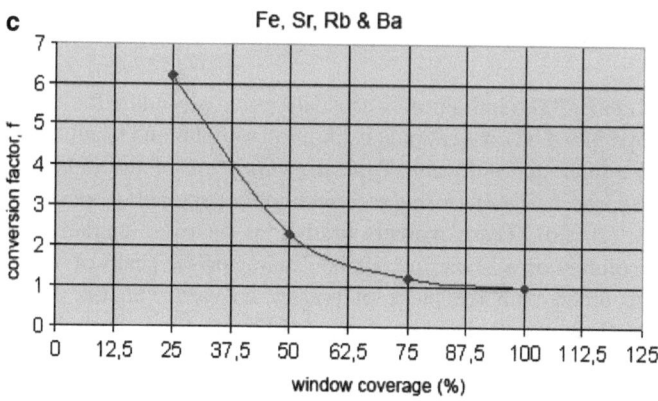

Fig. 6.8 Correction factors (*y*-axis) as a function of percentage coverage of window (*left* or *right*) for (**a**) K, (**b**) for Ca and Ti, and (**c**) Fe, Sr, Zr, Rb, Ba

program that simulated the X-ray spectrum from samples having a rough surface. The shape and intensity of such X-ray spectra were shown to be strongly influenced by changes in the generation and absorption of X-rays as the beam was moved across the

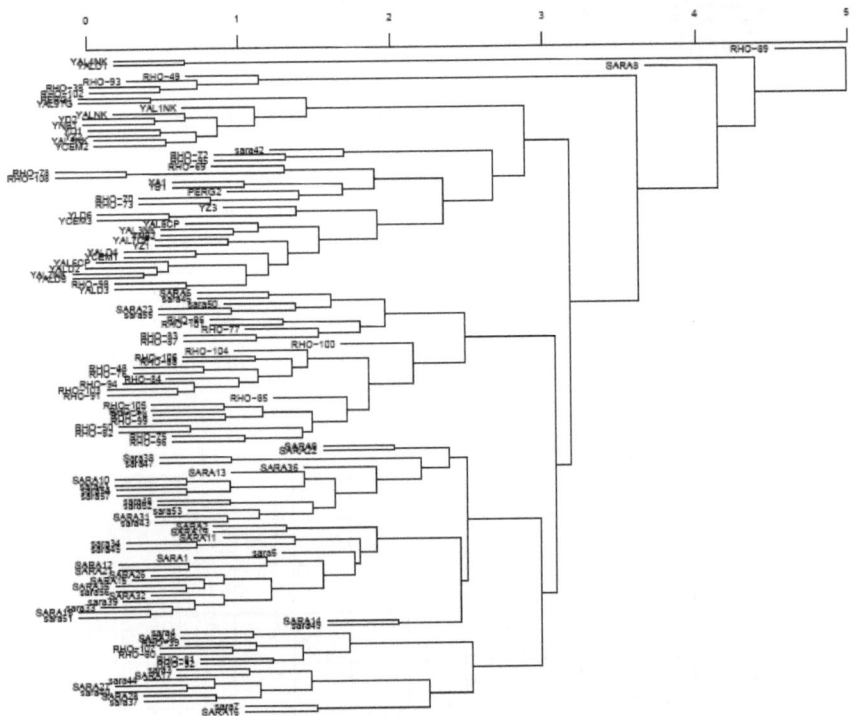

Fig. 6.9 A dendrogram of 188 ceramic and clay samples ranging from Neolithic to Bronze Age from Aegean, Cyprus, and Asia Minor of elements measured by PXRF. It is based on cluster analysis (hierarchical, average linkage). The main aim here was to compare Bayesian technique of Reversible Jump Markov Chain Monte Carlo with hierarchical clustering. They suggest the same number of main groups (except outliers/singletons) (Papageorgiou and Liritzis 2007)

sample. Bos et al. (2000) introduced a new calibration procedure for small samples of irregular shape based on conventional calibration samples and small modifications to existing procedures and software. Typical errors were of the order of 1.4–1.5%. A different approach to surface shape correction was applied to ancient ceramics by Leung et al. (2000b). These workers used Y as an outer marker by coating an appropriate solution on a Mylar membrane. Characteristic peaks of Y Kα and Y Kβ were then recorded with the piece of pottery covered with the membrane with correction coefficients calculated for elements of interest from test samples measured at discrete distances up to about 10 mm from the analyzer. (5) Detection limits near the limit threshold. Concentrations around this cut-off should be rejected. (6) Recording X-rays and converting to concentration: The processing of X-ray and gamma ray spectra poses its own particular problems. Stressed here is the tactic of using real peak shapes in the fitting function as there are typically large amounts of low-energy tails on peaks, which must be fitted accurately in order to get reliable peak area data. Similar findings on the importance of including peak distortions are reported by Kondrashov et al. (2000) who also recommended the use of least moduli rather than least squares

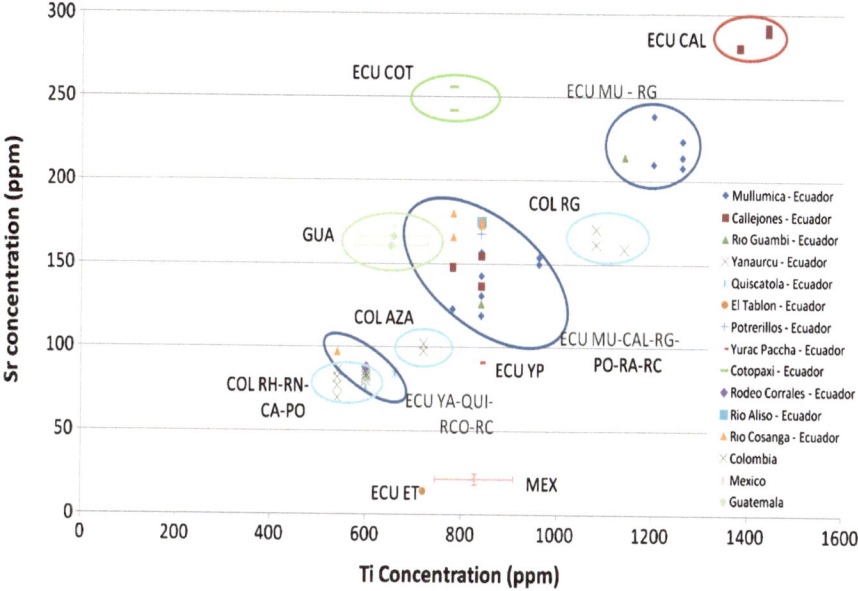

Fig. 6.10 Biplots of Sr vs. Ti for Ecuador and Colombia measured by ICP-AES, ICP-MS, and PIXE (Bellot-Gurlet et al. 2002, 2008), Mexico and Guatemala by PXRF (Liritzis 2007). Error *bars* are assigned when available, and ellipses are indicatively drawn. *ECU* Ecuador; *CAL* Callejones; *MU* Mullumica; *RG* Rio Guambi; *YA* Yanaurcu; *QUI* Quiscatola; *ET* El Tablon; *RCO* Rodeo Corrales; *RC* Rio Cosanga; *RA* Rio Aliso; *POT* Potrerillos; *COT* Cotopaxi; *YP* Yurac Paccha; *COL* Colombia; *RG* R10 Granates; *AZA* Azafatudo; *RH* R10 Hondo; *RN* R10 Negro; *CA* Caclites; *PO* Popayan; *MEX* Mexico; *GUA* Guatemala

method when fitting peaks with high peak-to-background ratio. It is well known that, no matter how good the peak fitting algorithm, the overall performance can only be as good as the detector response function, and while detector function is improved, peak fitting algorithms are crucial in reducing error. Many of the desktop manufacturers have been perfecting the peak fitting software for many years, but this has not necessarily translated to the PXRF instrumentation.

We suggest caution to users of ready built software in PXRF devices, and suggest interlaboratory comparison and proficiency tests, preferably with established laboratories (see for example IAEA Report IEAEA-CU-2006-06, CRP Project F.2.30.23, proficiency test on the determination of major, minor and trace elements in ancient Chinese ceramics).

Last but not least, the spectrum evaluation, matrix correction, and calibration procedures are most critical parameters, needing the attention of any PXRF user of applications in art and archaeometry, especially due to the wide range mineralogy of cultural materials studied.

Among the advantages are (1) the non-destructive/non-invasive procedure of measurement, (2) the portability (fieldwork, museums), (3) the fast processing, and (4) the satisfactory accuracy and precision in solving many archaeological issues.

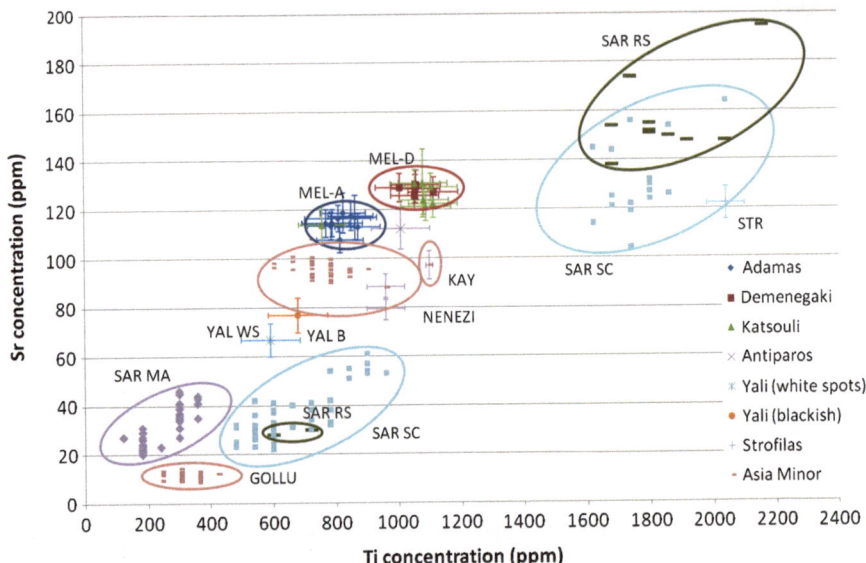

Fig. 6.11 Biplot of Sr vs. Ti. Aegean samples (Melos, Yali, Strofilas) were measured by portable EDXRF (Liritzis 2007), Asia Minor by ICP-MS, ICP-AES, LA-ICP-MS (Carter et al. 2006) and Sardinia by PIXE and SEM-EDS (Lugliè et al. 2006, 2007). Error *bars* are assigned when available, and ellipses are indicatively drawn. *MEL* Melos; *A* Adamas; *D* Demenegaki. Asia Minor, *GOLLU* Gollu Dag East; *NENEZI* Nenezi Dag; *KAY* Kayerli; *YAL* Yali; *WS* white spots; *B* blakish; *STR* Strofila; *ANT* Antiparos; *SAR* Sardinia; *RS* Rio Saboccu; *SC* Su Carroppu; *MA* Monte Arci

Conclusion

Chemical analysis by the portable WD- or EDXRF analyzers is efficient and accurate, and the equipment is versatile for in situ work and work in the laboratory. In the latter and window-type apertures, with appropriate calibration, smaller than the aperture size objects can be measured, using appropriately constructed set of curves per element, for which correction factors are devised per fractional coverage of the window. For other analyzers with focused X-rays beams particular surface points can be efficiently measured.

Grain size is essential for the evaluation of elemental contents, the fine grain size preferred and giving the highest values. If solid flat surfaces are used, the difference between those and the powdered samples is within counting errors, but not for some heavy elements such as Ti and Fe.

Sample thickness is essential in PXRF analysis of thin objects or unavailability of large quantities of powder. The critical depth Dc of the secondary X-rays depends upon the atomic number Z of the searched element, thus appropriate curves of Dc vs. Z may be produced for different rock types that assess this point.

The calibration as well as the interlaboratory comparison is essential .

PXRF instruments are extremely useful in a variety of applications in archaeology and history of art as they help to avoid destruction and conform to archaeologists' demand of non-invasive sampling. Many worldwide applications to cultural materials including ceramics, clays, soils, painted surfaces, and metals, enhance the value of PXRF and have produced useful results.

PXRF offers a unique opportunity for the non-destructive study of ancient materials, and portability is critical in making this technique effective. In this area, big steps forward have been taken: spectrometers weighing hardly a few kilograms have been built, thanks to the combined use of miniaturized X-ray tubes and thermoelectrically cooled detectors. Such advancement in portables rivals and, in most cases, acts independently of cumbersome laboratory instruments.

Acknowledgments We thank Prof G. Poupeau (Bordeaux) for the constructive correspondence and Dr. A. Vafiadou for her assistance during the preparation of the two biplots presented.

References

Acquafredda, P., Andriani, S., Lorenzoni, S., Zanettin, E., (1999) Chemical characterization of obsidians from different Mediterranean sources by nondestructive SEMEDS analytical method. *Journal of Archaeological Sciences* 26, 315–325.

Agnoli, A., Calliari, I., Mazzocchin, G.-A., (2007) Use of different spectroscopic techniques in the analysis of roman age wall paintings. *Annali di Chimica* 97 (1–2), 1–7.

AlKofahi, M.M., AlTarawneh, K.F., (2000) Analysis of Ayyubid and Mamluk dirhams using X ray fluorescence spectrometry. *X-Ray Spectrometry* 29, 39–47.

Andrikopoulos, K.S., Daniilia, S., Roussel, B., Janssens, K., (2006) In vitro validation of a mobile Raman-XRF microanalytical instrument's capabilities on the diagnosis of Byzantine icons. *Journal of Raman Spectroscopy* 37 (10), 1026–1034.

Appoloni, C.R., Blonski, M.S., Parreira, P.S., Souza, L.A.C., (2007) Pigments elementary chemical composition study of a gainsborough attributed painting employing a portable Xrays fluorescence system. *AIP Conference Proceedings* 884, 459–464.

Baxter, M.J., (1994) *Exploratory multivariate analysis in archaeology.* Edinburgh University Press, Scotland.

BellotGurlet, L., Le Bourdonnec, F.X., Poupeau, G., Dubernet, S., Bos, M., Vrielink, J.A.M., van der Linden, W.E., (2000) Nondestructive analysis of small irregularly shaped homogenous samples by Xray fluorescence spectrometry. *Analytica Chimica Acta* 412, 203–211.

BellotGurlet, L., Dorighel, O., Poupeau, G., Keller, F., Scorzelli, R.B., (2002) First characterization of obsidian from Colombian and Ecuadorian sources using ICPAES and ICPMS. *Proceedings of the 31st International Symposium on Archaeometry*, Jerem, E. and Biro, T. (eds.), Archaeopress Archaeolingua, BAR International Series 1043 (II), 678–684.

BellotGurlet, L., Dorighel, O., Poupeau, G., (2008) Obsidian provenance studies in Colombia and Ecuador: obsidian sources revisited. *Journal of Archaeological Science* 35 (2), 272–289.

Bos, M., Vrielink, J.A.M., Van der Linden, W.E., (2000) Nondestructive analysis of small irregularly shaped homogeneous samples by X ray fluorescence spectrometry. *Analytica Chimica Acta* 412, 203–211.

Buxeda i Carrigos, J., Jones, R.E., Kilikoglou, V., Levi, S.T., Maniatis, Y., Mitchell, J., Vagretti, L., Wardle, K.A., Andrews, S., (2003) Technology transfer at the periphery of the Mycenaean World: the cases of Mycenaen pottery found in Central Macedonia (Greece) and the plain of Sybaris (Italy). *Archaeometry* 45 (2), 263–284.

Capitan - Vallvey, L.F., Manzano, E., Medina - Florez, V.J., (1994) A study of the materials in the mural paintings at the "Corral del Carbon" in Granada, Spain. *Studies in Conservation* 39 (2), 87–99.

Carter, T., Poupeau, G., Bressy, C., Pearce, N., (2006) A new programme of obsidian characterization at Catalhoyuk, Turkey. *Journal of Archaeological Sciences* 33 (7), 893–909.

Cesareo, R., Gigante, G.E., Iwanczyk, J.S., Dabrowski, A., (1992) Use of a mercury iodide detector for Xray fluorescence analysis in archaeometry. *Nuclear Instruments and Methods in Physics Research A* 322, 583–590.

Cesareo, R., Gigante, G.E., Canegallo, P., Castellano, A., Iwanczyk, J.S., Dabrowski, A., (1996) Applications of noncryogenic portable EDXRF systems in archaeometry. *Nuclear Instruments and Methods* 380, 440–445.

Cesareo, R., Gigante, G.E., Castellano, A., (1999) Thermoelectrically cooled semiconductor detectors for nondestructive analysis of works of art by means of energy dispersive Xray fluorescence. *Nuclear Instruments and Methods A* 428, 171–181.

Cesareo, R., Cappio Borlino, C., Stara, G., Brunetti, A., Castellano, A., Buccolieri, G., Marabelli, M., Giovagnoli, A.M., Gorghinian, A., Gigante, G.E., (2000) A portable EDXRF apparatus for the analysis of sulphur and chlorine in frescoes and stony monuments. *Trace Microprobe Technology* 18 (1), 23–33.

Cojocaru, V., Constantinescu, B., Stefanescu, I., Petolescu, C.M., (2000) EDXRF and PAA analyses of Dacian gold coins of "Koson" type. *Journal of Radioanalytical and Nuclear Chemistry* 246 (1), 185–190.

Craig, N., Speakman, R.J., PopelkaFilcoff, R.S., Glascock, M.D., Robertson, J.D., Shackley, M.S., Aldenderfer, M.S., (2007) Comparison of XRF and PXRF for analysis of archaeological obsidian from southern Perú. *Journal of Archaeological Science* 34 (12), 2012–2024.

Davis, M.K., Jackson, T.L., Shackley, M.S., Teague, T., Hampel, J.H., (1998) Factors affecting the energy dispersive X ray fluorescence (EDXRF) analysis of archaeological obsidian. In Shackley, M.S. (ed.), *Archaeological Obsidian studies, method and theory*, Plenum Press, New York, 159–180.

De Fransesco, A.M., Crisci, G.M., Bocci, M., (2007) Non destructive analytical method using XRF for determination of provenance of archaeological obsidian from the Mediterranean area: a comparison with traditional XRF method. *Archaeometry* 50 (2), 337–350.

Desnica, V., Škarić, K., JembrihSimbuerger, D., Fazinić, S., Jakšić, M., Mudronja, D., Pavličić,M., Peranić, I., Schreiner, M., (2008). Portable XRF as a valuable device for preliminary in situ pigment investigation of wooden inventory in the Trski Vrh Church in Croatia. *Applied Physics A: Materials Science and Processing* 92 (1), 19–23.

Ebel, H., (1999) Xray tube spectra. *X-Ray Spectrometry* 28, 255–266.

Ferretti, M., Moioli, P., (1998) The use of portable XRF systems for preliminary compositional surveys on large bronze objects. A critical review after some years' experience. In *Proceedings of the International Conference Metal 98*, Draguignan 2729 May 1998, Mourey, W. and Robiola, L. (eds.), 39–44.

Ferretti, M., Guidi, G., Moioli, P., Scafe R., Seccaroni C., (1991) The presence of antimony in some grey colours of three paintings by Correggio. *Studies in Conservation* 36, 235–239.

Ferretti, M., Miazzo, L., Moioli, P., (1997) The application of a nondestructive XRF method to identify different alloys in the bronze statue of the Capitoline Horse. *Studies in Conservation* 42, 241–246.

Gauvin, R., Lifshin, E., (2000) Simulation of X ray emission from rough surfaces. *Mikrochimica Acta* 132, 201–204.

Gopher, Z., (1983) Physical studies of archaeological materials. *Report Progress on Physics* 46, 1193–1234.

Guerra, M.F., (2008). An overview on the ancient goldsmith's skill and the circulation of gold in the past: the role of Xray based techniques. *X-Ray Spectrometry* 37 (4), 317–327.

Hall, E.T., Schweizer, F., Toller, P.A., (1973) Xray fluorescence analysis of museum objects: a new instrument. *Archaeometry* 15, 53–78.

Haruyama, Y., Saito, M., Muneda, T., Mitani, M., Yamamoto, R., Yoshida, K., (1999).Comparison between PIXE and XRF for old Japanese copper coin analysis. *International Journal of PIXE* 9, 181–188.

Kallithrakas-Kontos, N., Katsanos, A.A., Touratsoglou, J., (2000) Trace element analysis of Alexander the great's silver tetradrachms minted in Macedonia. *Nuclear Instruments and Methods in Physics Research B* 171 (3), 342–349.

Karydas, A.G., (2007) Application of a portable XRF spectrometer for the noninvasive analysis of museum metal artefacts. *Annali di Chimica* 97 (7), 419–432.

Karydas, A.G., Kotzamani, D., Bernard, R., Barrandon, J.W., Zarkadas, Ch., (2004) A compositional study of a museum jewellery collection (7th - 1st c. BC) by means of a portable XRF spectrometer. Nucl. Instr. Meth. in Physics Res. B 226, 15–28.

Kitov, B.I., (2000) Calculation features of the fundamental parameter method in XRF. *X-Ray Spectrometry* 29, 285–290.

Knoll, G.F., (2000) Radiation detectors for Xray and gammaray spectroscopy. *Journal of Radioanalytical and Nuclear Chemistry* 243 (1), 125–131.

Kondrashov, V.S., Rothenberg, S.J., SajoBohus, L., Greaves, E.D., Liendo, J.A., (2000) Increasing reliability in gamma and X-ray spectral analysis: least moduli approach. *Nuclear Instruments and Methods in Physics Research A* 446 (3), 560–568.

Kunicki-Goldfinger, J., Kierzek, J., Kasprzak, A., Malozewska - Bucko, B., (2000) A study of eighteenth century glass vessels from central Europe by xray fluorescence analysis. *X-Ray Spectrometry* 29, 310–316.

Langhoff, N., Arkadiev, V.A., Bjeoumikhov, A.A., Gorny, H.E., Schmalz, J., Wedell, R., (1999) Concepts for a portable X-ray spectrometer for nondestructive analysis of works of art. *Berliner Beiträge zur Archäometrie* 16, 155–161.

Leslie, C., Matthew, G., Moriarty, D., Speakman, R.J., Glascock, M.D., (2007) Feasibility of field portable XRF to identify obsidian sources in Central Petén, Guatemala. In *Archaeological chemistry: analytical methods and archaeological interpretation*, Glascock, M.D., Speakman, R.J.,Popelka Filcoff, R.S. (eds.), 506–521. ACS Publication Series 968. American Chemical Society, Washington, DC.

Leung, P.L., Daze, S., Stokes, M.J., (2000a) EDXRF surface shape correction for thick sample measurement using an outer mark membrane. *X-Ray Spectrometry* 29 (5), 360–364.

Leung, P.L., Peng, Z.C., Stokes, M.J., Li, M.T.W., (2000b) EDXRF studies of porcelains (8001600 A.D.) from Fujian, China with chemical proxies and principal component analysis. *X-Ray Spectrometry* 29(5), 253–259.

Linke, R., Schreiner, M., (2000) Energy dispersive Xray fluorescence analysis and Xray microanalysis of medieval silver coins. *Mikrochimica Acta* 133, 165–170.

Liritzis, I., (2005) Ulucak (Smyrna, Turkey): chemical analysis with clustering of ceramics and soils and obsidian hydration dating. *Mediterranean Archaeology and Archaeometry* 5(3), Special Issue, 33–45.

Liritzis, I., (2007) Assessment of Aegean obsidian sources by a portable EDXRF analyzer (grouping, provenance and accuracy). In *Proceedings of the 4th Symposium of the Hellenic Society for Archaeometry*, Facorellis, Y., Zacharias, N., Polikreti, K. (eds.), Archaeopress, BAR International Series 1746, 399–406.

Liritzis, I., Polychroniadou, E., (2007) Optical and analytical techniques applied to the Amfissa Cathedral mural paintings made by the Greek artist Spyros Papaloukas (1892–1957). *Revue d' Archaeometrie (Archaeosciences)* 31, 97–112.

Liritzis, I., Drakonaki, S., Vafiadou, A., Sampson, A., Boutsika, T., (2002) Destructive and nondestructive analysis of ceramics, artefacts and sediments of Neolithic Ftelia (Mykonos) by portable EDXRF spectrometer: first results. In Sampson, A. (ed.), *The Neolithic settlement at Ftelia, Mykonos*, University of the Aegean, Department of Mediterranean Studies, Rhodes, 251–272.

Liritzis, I., Sideris, C., Vafiadou, A., Mitsis, J., (2007) Mineralogical petrological and radioactivity aspects of some building material from Egyptian Old Kingdom monuments. *Journal of Cultural Heritage* 9, 1–13.

Longoni, A., Fiorini, C., Leutenegger, P., Sciuti, S., Fonterotta, G., Stróder, L., Lechner, P., (1998) A portable XRF spectrometer for nondestructive analyses in archaeometry. *Nuclear Instruments and Methods A* 409, 407–409.

Lugliè, C., Le Bourdonnec, F.X., Poupeau, G., Bohn, M., Meloni, S., Oddone M., Tanda, G., (2006) A map of the Monte Arci (Sardinia Island, Western Mediterranean) obsidian primary to secondary sources. Implications for Neolithic provenance studies. *C R Paleo* 5, 995–1003.

Lugliè, C., Le Bourdonnec, F.X., Poupeau, G., Atzeni, E., Dubernet, S., Moretto P., Serani, L., (2007). Early Neolithic obsidians in Sardinia (Western Mediterranean): the Su Carroppu case. *Journal of Archaeological Science* 34, 428–439.

Mantzourani, H., Liritzis, I., (2006) Chemical analysis of pottery samples from Kantou Kouphovounos and Sotira Tepes (Cyprus): a comparative approach. Reports of the Department of Antiquities, Cyprus, 63–76.

Papadopoulou, D.N., Zachariadis, G.A., Anthemidis, A.N., Tsirliganis, N.C., Stratis, J.A., (2006). Development and optimisation of a portable microXRF method for in situ multielement analysis of ancient ceramics. *Talanta* 68 (5), 1692–1699.

Papadopoulou, D., Sakalis, A., Merousis, N., Tsirliganis, N.C., (2007). Study of decorated archaeological ceramics by micro Xray fluorescence spectroscopy. *Nuclear Instruments and Methods in Physics Research A* 580 (1), 743–746.

Papageorgiou, I., Liritzis, I., (2007) Multivariate mixture of normals with unknown number of components. An application to cluster Neolithic ceramics from the Aegean and Asia Minor. *Archaeometry* 49 (4), 795–813.

Pappalardo, G., Karydas, A.G., La Rosa, V., Militello, P., Pappalardo, L., Rizzo, F., Romana, F.P., (2003) Provenance of obsidian artefacts from different archaeological layers of Phaistos and Hagia Triada. *Creta Antica* 4, 287–300.

Pappalardo, L., Karydas, A.G., Kotzamani, N., Pappalardo, G., Romano, F.P., Zarkadas, Ch., (2005). Complementary use of PIXE-alpha and XRF portable systems for the nondestructive and in situ characterization of gemstones in museums. *Nuclear Instruments and Methods in Physics Research B* 239 (12), 114–121.

Pérez-Arantegui, J., Resano, M., García - Ruiz, E., Vanhaecke, F., Roldán, C., Ferrero, J., Coll, J., (2008). Characterization of cobalt pigments found in traditional Valencian ceramics by means of laser ablation inductively coupled plasma mass spectrometry and portable Xray fluorescence spectrometry. *Talanta* 74 (5), 1271–1280.

Pillay, A.E., (2001) Analysis of archaeological artefacts: PIXE, XRF or ICPMS?. *Journal of Radioanalytical and Nuclear Chemistry* 247 (3), 593–595.

Pollard, A. M., (1986) Multivariate methods of data analysi. In Greek and Cypriot pottery: a review of scientific studies, (ed. R. E. Jones). Fitch Lab. Occas. Pap., 1, Brit. Sch. Athens, 56–83, Athens.

Potts, J.P., West, M., (eds), (2008). *Portable Xray fluorescence spectrometry: capabilities for in situ analysis*. The Royal Society of Chemistry, Cambridge.

Potts, J.P., Webb, P.C., Williams - Thorpe, O., (1995) Analysis of silicate rocks using fieldportable Xray fluorescence instrumentation incorporating a mercury (II) iodide detector: a preliminary assessment of analytical performance. *Analyst* 120, 1273–1278.

Potts, J.P., Williams-Thorpe, O., Webb, C.P., (1997) The bulk analysis of silicate rocks by portable XRay fluorescence: Effect of sample mineralogy in relation to the size of the excited volume, Geostandards Newsletter. *The Journal of Geostandards and Geoanalysis* 21 (1), 29–41.

Potts, J.P., Ellis, A.T., Kregsamer, P., Marshall, J., Streli, C., West, M., Wobrauschek, P., (2001) Atomic spectrometry update: Xray fluorescence spectrometry (The Royal Society of Chemistry). *Journal of Analytical and Atomic Spectrometry* 16, 1217–1237.

Rebocho, J., Carvalho, M.L., Marques, A.F., Ferreira, F.R., Chettle, D.R., (2006) Lead postmortem intake in human bones of ancient populations by 109Cd based Xray fluorescence and EDXRF. *Talanta* 70 (5), 957–961.

Romano, F.P., Pappalardo, G., Pappalardo, L., Garraffo, S., Gigli, R., Pautasso, A., (2006) Quantitative nondestructive determination of trace elements in archaeological pottery using a portable beam stability controlled XRF spectrometer. *X-Ray Spectrometry* 35 (1), 17.

Roth, B.J., (2000) Obsidian source characterization and hunter gatherer mobility: an example from the Tuscon basin. *Journal of Archaeological Science* 27, 305–314.

Rotondi, G., Urbani, G., (1972) Non destructive analysis of chemical elements in paintings and enamels. *Archaeometry* 14, 65–78.

Sandor, Z., Tolgyesi, S., Gresits, I., Kaplan - Juhasz, M., (2000) Qualitative and quantitative analysis of medieval silver coins by energy dispersive Xray fluorescence method. *Journal of Radioanalytical Nuclear Chemistry* 246 (2), 385–389.

Schwedt, A., Mommsen, H., Zacharias, N., (2004) Postdepositional elemental alterations in pottery: neutron activation analyses of surface and core samples. *Archaeometry* 46, 85–101.

Shackley, M.S., (2005) *Obsidian. Geology and archaeology in the North American Southwest.* University of Arizona Press, Tucson.

Sokaras, D., Karydas, A.G., Oikonomou, A., Zacharias, N., Beltsios, K., Kantarelou, V., (2009) Combined elemental analysis of ancient glass beads by means of ion-beam, portable XRF and EPMA techniques, *Analytical Bioanalytical Chemistry* 395, 199–2209.

Spoto, G., Torrisi, A., Contino, A., (2000) Probing archaeological and artistic solid materials by spatially resolved analytical techniques. *Chemical Society Reviews* 29 (6), 429.

Tite, M.S., (2008) Ceramic production, provenance and use: a review. *Archaeometry* 50, 216–231.

Tykot, R., (2001) Chemical fingerprint and source tracing of obsidian: the central Mediterranean trade in black gold. *Accounts of Chemical Research* 35, 618–627.

Uda, M., Sassa, S., Taniguchi, K., Nomura, S., Yoshimura, S., Kondo, J., Iskander, N., Zaghloul, B., (2000) Touchfree in situ investigation of ancient Egyptian pigments. *Naturwissenschaften* 87 (6), 260–263

Uda, M., Demortier, G., Nakai, I., (eds.), (2005) *X rays in archaeology.* The Netherlands, Springer.

Vandenabeele, P., Moens, L., Edwards, H.G.M., Dams, R., (2000a) Raman spectroscopic database of Azopigments and application to modern art studies. *Journal of Raman Spectroscopy* 31 (6), 509–517.

Vandenabeele, P., Wehling, B., Moens, L., Edwards, H., De Reu, M., Van Hooydonk, G., (2000b) Analysis with microRaman spectroscopy of natural organic binding media and varnishes used in art. *Analytical Chimica Acta* 407, 261–274.

Wegrzynek, D., (2005) (Trans) portable XRF spectrometer with polycapillary optics and vacuum chamber. XRF Newsletter, IAEA, Issue 10, December.

Williams-Thorpe, O., (1995) Obsidian in the Mediterranean and Near East: a provenancing success story. *Archaeometry* 37, 217–248.

Williams-Thorpe O., Potts, P.J., Webb, P.C., (1999) Field portable non destructive analysis of lithic archaeological samples by X ray fluorescence instrumentation using a mercury iodide detector: comparison with wavelength – dispersive XRF and a case study in British stone Axe provenancing. *Journal of Archaeological Science* 26 (2), 215–237.

Willis, J.P., Lachance, G.R., (2000) Resolving apparent differences in mathematical expressions relating intensity to concentration in Xray fluorescence spectrometry. *The Rigaku Journal* 17 (1), 23–33.

Willis, J.P., Lachance, G.R., (2002) Debate on some algorithms relating concentration to intensity in XRF spectrometry. *The Rigaku Journal* 19 (1), 25–34.

Wobrauschek, P., Halmetschlager, G., Zamini, S., Jokubonis, C., Trnka, G., Karwowski, M., (2000) Energy dispersive xray fluorescence analysis of Celtic glasses. *X-Ray Spectrometry* 29, 25–33.

Wu, J., Leung, P.L., Li, J.Z., Stokes, M.J., Li, M.T.W., (2000) EDXRF studies on blue and white
    Chinese Jingdezhen porcelain samples from the Yuan, Ming and Qing dynasties. *X-Ray
    Spectrometry* 29, 239–244.
Zacharias, N., Beltsios, K., Oikonomou, Ar., Karydas, A.G., Bassiakos, Y., (2008) Thermally and
    optically stimulated luminescence properties of an archaeological glass collection from
    Thebes, Greece. *Journal of Non Crystalline Solids* 354, 761–767.
Zacharias, N., Bassiakos, Y., Hayden, B., Theodorakopoulou, K., Michael, C.T., (2009) Lumines-
    cence dating of deltaic deposits from eastern Crete, Greece: geoarchaeological implications.
    Geomorphology 109 (1–2), 46–53.

Chapter 7
Elemental Analysis of Fine-Grained Basalt Sources from the Samoan Island of Tutuila: Applications of Energy Dispersive X-Ray Fluorescence (EDXRF) and Instrumental Neutron Activation Analysis (INAA) Toward an Intra-Island Provenance Study

Phillip R. Johnson

Introduction

The following chapter presents the results from recent applications of energy disper-sive X-ray fluorescence (EDXRF) in the provenance study of fine-grained basalt procurement and production sites from the island of Tutuila, American Samoa. This research was designed to address two primary objectives. The first objective was the differentiation of four precontact fine-grained basalt procurement and manufacture sites, using elemental compositional data derived from EDXRF analysis. The second objective of the project was to evaluate the efficacy of EDXRF in the differentiation of those sites when compared against previous differentiation of the same sites (Johnson et al. 2007) using instrumental neutron activation analysis (INAA). Both XRF and INAA are widely established techniques for archaeometric provenance analyses (Bishop et al. 1990; Glascock 1992; Green 1998; Neff 2000; Shackley 1998a; see Chap. 8), but XRF is the technique of choice for the provenance analysis of Polynesia basalt artifacts and sources (Best et al. 1992; Clark et al. 1997; Kahn 2005; Lebo and Johnson 2007; Mills et al. 2008; Sheppard et al. 1997; Weisler 1993a, b, 1997, 1998; Winterhoff et al. 2007). Although XRF is the most commonly utilized technique in Polynesia provenance studies, INAA was previously selected by the author for the differentiation of Tutuila basalt sources (Johnson 2005; Johnson et al. 2007) because Clark et al. (1997, p. 81) reported difficulty differentiating between multiple intra-island Tutuila basalt sources (including those selected for this project) through XRF compositional data. The application of INAA at the Texas A&M EAL is thus far the only use of INAA toward archaeometric analysis of basalt sources in

P.R. Johnson (✉)
Department of Anthropology, Texas A&M University, College Station, TX 77843-4352, USA
e-mail: phillipjohnson@tamu.edu

M.S. Shackley (ed.), *X-Ray Fluorescence Spectrometry (XRF) in Geoarchaeology*,
DOI 10.1007/978-1-4419-6886-9_7, © Springer Science+Business Media, LLC 2011

West Polynesia and has resulted in differentiation of multiple intra-island basalt procurement sites (Johnson et al. 2007).

Although differentiation of intra-island fine-grained basalt procurement sites was achieved using INAA, there were several factors that led to this application of EDXRF for the analysis of Tutuila basalt sources and production sites. The first factor was the aforementioned preference, frequency and success for XRF analysis in the archaeometric provenance study of Polynesian basalt artifacts and sources. The second factor was that sample preparation and analysis for EDXRF is less time consuming and destructive than sample preparation and analysis for INAA. The quicker turnaround in both the preparation and analysis of samples makes EDXRF attractive, especially when analyzing hundreds of samples. In addition to quicker turnaround, the ability to perform possible nondestructive analyses of artifacts using EDXRF is especially compelling when dealing with culturally sensitive materials that may otherwise not be available for destructive analysis (Mills et al. 2008). The final factor leading to this research was the successful differentiation of several Tutuila basalt tool production sites by Winterhoff et al. (2007) using wavelength dispersive X-ray fluorescence (WDXRF). This successful WDXRF characterization of multiple basalt tool production areas located in a single valley was compelling support for the possibility of differentiating intra-island sources using EDXRF.

Geography and Geology of the Research Area

The Samoan archipelago is comprised of nine major islands formed by oceanic basalt shield volcanoes that trend easterly (MacDougall 1985). The West Polynesian island chain lies east of the andesite line (Fig. 7.1), a petrographic boundary that splits the South Pacific. Samoan shield building volcanism, comprised primarily of alkalic olivine basalts and hawaiities (MacDonald 1968), began several million years ago and ceased approximately one million years ago (MacDougall 1985). The western-most islands are the oldest and the Manu'a islands in the east are the youngest, but while shield building activity trended to the east, post-erosional volcanism trended westerly (MacDougall 1985; Natland 1980). The island of Tutuila lies in the center of the Samoan archipelago at approximately 14° South Latitude and 170°E Longitude (Fig. 7.1). The third largest of the Samoan islands, Tutuila is a narrow mountainous landform approximately 138 km^2 in total area.

One of the earliest published commentaries on the geology of the Samoan islands was presented in a missive to the Honolulu based newspaper *The Polynesian* by the missionary Heath, dated Saturday September 19, 1840. In his observations on the geological composition and diversity of the largest Samoan islands Heath said, "It has been stated that the surface of this group is almost entirely volcanic, so that the geologist will not find much variety. At Tutuila, however, is found the hard stone (Trap,) of which the Polynesian adzes and other tools were made previously to the introduction of iron. At the other islands the stone is almost uniformly porous

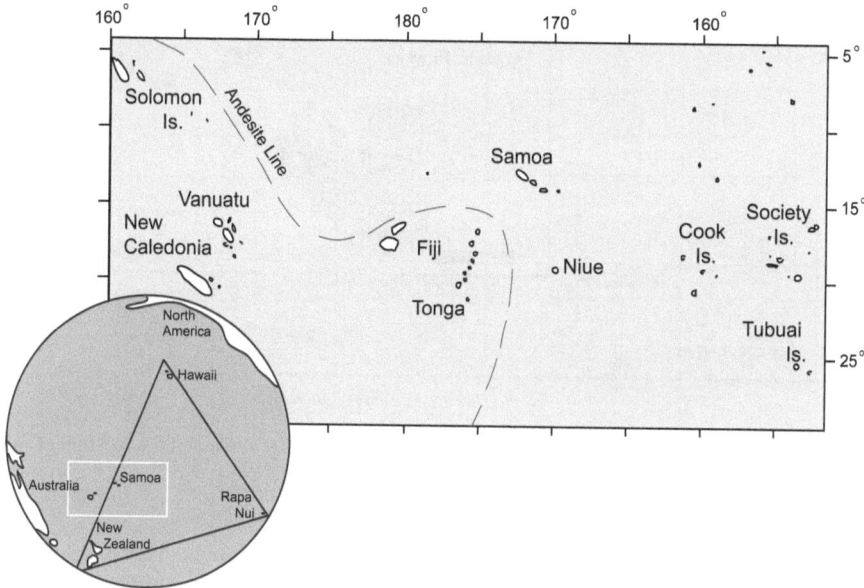

Fig. 7.1 Map of the Polynesian Triangle and the islands of West Polynesia

and of a dull black color (Heath 1840)." Over 100 years after the observations of Heath, the Bulletin of the Geological Society of America published the foundational geologic survey and descriptions of Tutuila by Stearns (1944) along with the petrography of MacDonald (1944). Stearns (1944) defined five major volcanic provinces for Tutuila; the four essentially contemporaneous westward expanding shield volcanic centers Olomoana, Alofau, Pago, Taputapu, and the post-erosional Leone Volcanics (Fig. 7.2). Although Stearns' (1944) work remains a primary resource, recent research has added to the understanding of the island's formation (MacDougall 1985; Natland 1980, 2003; Wright 1986). MacDougall (1985) performed Ka-Ar dating that supports Stearns' (1944) chronology of contemporaneous shield building activity, but argues that the Alofau volcanics are actually contained within the eastern flank of the Pago volcanics. Sampling and analysis for this project were based primarily on Stearns' (1944) original interpretations but employed the interpretation of MacDougall (1985) and included the Alofau volcanics within the Pago volcanic province (Fig. 7.2).

Archaeological Context

There are no less than 20 recorded fine-grained basalt procurement and production sites on the island of Tutuila (Clark et al. 1997; Johnson et al. 2007; Winterhoff et al. 2007). The majority of recorded basalt procurement and production sites on the island have been briefly described (Clark 1989; Clark et al. 1997), but the

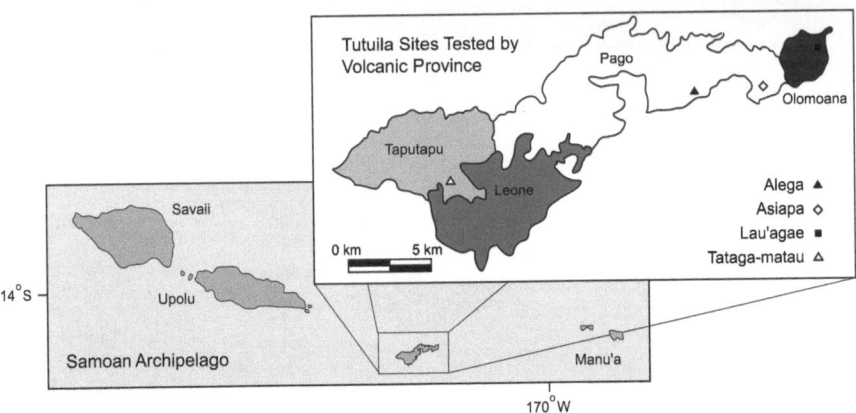

Fig. 7.2 Map of the Samoan Archipelago including the island of Tutuila and the location of the fine-grained basalt procurement sites included in this research

sites of Alega (Clark 1992), Lau'agae (Moore and Kennedy 1996), Maloata (Winterhoff 2007), and Tataga-matau (Leach and Witter 1987, 1990) have been the focus of more detailed discussions. This research sampled four previously recorded and characterized procurement sites from each shield volcanic province (Fig. 7.2). The four sites included in this analysis were Alega ($n = 18$), Asiapa ($n = 18$), Lau'agae ($n = 18$), and Tataga-matau ($n = 18$). All four sites have been included in previous chemical characterization projects (Best et al. 1992; Clark et al. 1997; Johnson et al. 2007; Moore and Kennedy 1996). The sites of Alega and Asiapa were sampled from the Pago volcanics, samples from the site of Lau'agae in the Olomoana province were selected, and samples from Tataga-matau were selected to represent the Taputapu volcanics. There is currently no recorded basalt procurement site located in the Leone volcanics, which at the surface is comprised largely of post-erosional vesicular basalt, and no samples were selected from this area.

Clark (1992) reported three areas of fine-grained basalt procurement and tool production above the modern village of Alega that he labeled Alega 1 (AS-23-22), Alega 2 (AS-23-22), and Alega 3 (AS-23-29). All samples for this research were collected from Alega 2 because modern industrial activity has destroyed the remnants of Alega 1 and Alega 3 (Johnson 2005). Asiapa (AS-22-31) is a site on the southeastern ridge of Asiapa mountain in the eastern flank of the Pago volcanics in the area that Stearns (1944) had previously identified as the Alofau volcanic province. During the exploratory surveys of the East Tutuila Project, Clark (1989) reported lithic scatters at the site that covered an area of approximately 205 m^2. The site known as the Lau'agae quarry (AS-21-100) is located on Cape Matatula in the eastern province of the Olomoana volcanics. Along with Alega and Asiapa, this site was discovered during the survey of the East Tutuila Project (Clark 1989). Moore and Kennedy (1996) reported that the site consisted of no less than 12 discrete areas of basalt procurement and stone tool manufacture totaling approximately 10,000 m^2.

Tataga-matau (AS-34-10), located in the Taputapu Volcanics, is the most cele-brated and investigated archaeological site on Tutuila, if not the entire Samoan archipelago. Investigation of this site began with Sir Peter Buck (Te Rangi Hiroa) in 1927 (Buck 1930), but it was not again investigated until Kikuchi (1963) and Clark (1980) revisited the Leone Valley. Tataga-matau was the subject of multiple investigations in the 1980s by Leach and Witter (1985, 1987, 1990) and Best et al. (1989). The site is described as a complex system of surface features including, but not limited to, fortifications, mounds, pits, terraces, and three distinct basalt procurement and lithic manufacture areas (Best et al. 1989). Tataga-matau has also featured very prominently over any other Tutuila basalt procurement and tool manufacture site in the investigation of long-distance interaction and exchange (Best et al. 1992; Clark et al. 1997; Weisler and Kirch 1996).

Regional Chemical Characterization Studies

The island societies of Polynesia (Fig. 7.1) were established in the late Holocene through multiple long distance ocean voyages (Kirch and Green 2001) and maintained through inter-island and inter-archipelago maritime contact (Davidson 1977; Kaep-pler 1978; Weisler 1998). This Polynesian diaspora and continued long-distance interaction have been a primary impetus for archaeological investigation, and the subsequent use of provenance analyses for investigation of ocean voyaging and interaction in the region. Elemental analysis of lithic artifacts and their material sources has a long standing position in Polynesia archaeology, beginning with the early research of Roger Green (1962, 1964) on obsidian artifacts and sources and eventually the application of geochemical provenance analysis on basalt artifacts and sources (e.g., Parker and Sheppard 1997; Weisler 1990, 1993b, 2003; Weisler and Sinton 1997; Weisler and Woodhead 1995). Over the last 2 decades, basalt artifacts have become the focus for the majority of geochemical provenance studies in Polynesia due to a dearth of pottery and volcanic glass or obsidian sources throughout the region. Most often basalt geochemical provenance analysis in Poly-nesia has been used in the investigation of inter-island exchange (Collerson and Weisler 2007; Rolett et al. 1997; Sheppard et al. 1997; Weisler 1997, 1998, 2002; Weisler and Kirch 1996; Weisler et al. 1994). The investigation of long-distance interaction has established the Samoan island of Tutuila as a significant source for fine-grained basalt throughout West Polynesia and across the South Pacific (Allen and Johnson 1997; Best et al. 1992; Clark et al. 1997; Weisler 1993a; Winterhoff 2007), and this evidence for the long-distance exchange of basalt artifacts has featured prominently in most geochemical provenance studies involving Tutuila. Although there are over 20 known basalt manufacture and production sites on Tutuila, very few projects have focused primarily on the differentiation of multiple intra-island sources and artifacts (Clark et al. 1997; Crews 2008; Johnson et al. 2007; Winterhoff et al. 2007). This project was designed as an addition to the growing body of research toward the characterization of intra-island fine-grained basalt source variability on Tutuila.

EDXRF: Materials and Methods

Sample preparation and EDXRF analysis for this project were conducted by the author at the Elemental Analysis Laboratory (EAL) in the Texas A&M University Center for Chemical Characterization. The EAL has been conducting archaeometric analyses for Texas A&M and outside patrons for nearly 2 decades, but the majority of those projects have utilized INAA (James et al. 2007). This research represents the first application of quantitative EDXRF for an archaeometric provenance study at the EAL. All analyses for this project were conducted on the EALs Thermo *QuantX* EC EDXRF spectrometer equipped with a liquid nitrogen cooled Si(Li) detector. The spectrometer was calibrated for quantitative analysis using pure-element reference spectra and powdered geological standards from the United States Geological Survey (USGS) and the National Institute of Standards and Technology (NIST). A total of nine USGS standards (AGV-1, BCR-2, BHVO-1, BHV0-2, BIR-1a, DNC-1, GSP-2, QLO-1, W-2) and one NIST standard (SRM-688) were used in this calibration. All geologic standards used for calibration and control as well as the basalt samples from Tutuila were pressed into approximately 4 g pellets. Eighteen samples from four separate sites were included in this research for a total number of 72 samples analyzed. All samples were collected in the field by the author in 2004, and the design for the original field sampling is detailed in Johnson et al. (2007). Basalt samples were selected for this project from reserved material previously collected for INAA (Johnson et al. 2007) and curated at the Texas A&M University Anthropology Department.

It was necessary to analyze pressed pellets because the majority of samples held in reserve from the previous INAA project were crushed internal fragments that were determined to be too small for direct (i.e., nondestructive) application of EDXRF (Lunblad et al. 2008). For this project, pellets were prepared by combining approximately 0.5 ml of a 3% solution of polyvinyl alcohol (PVA) binder with 4 g of powdered rock material in a methylacrylate vial and ball set and then agitated for 5 min in a Spex Certiprep 8000 Mixer/Mill. After agitation, the powder/PVA mixture was pressed into pellets using a Spex Certiprep 25-ton laboratory press. After pressing, the pellets were dried in a 110° oven for 3 h. During the analysis, the USGS standard BHVO-2 and the NIST standard SRM 688 were included with the basalt samples as a control and repeatedly measured. EDXRF analytical conditions selected for the analysis for the Low-Za, Mid-Za, and Mid-Zc elements as designated in the WinTrace™ software were derived directly from the Polynesia basalt-specific methodology established at the University of Hawaii at Hilo by Lunblad et al. (2008, p. 4; see also Chap. 4).

Results

This research reports concentrations for 15 elements attained through EDXRF analysis of basalt samples from Tutuila. Table 7.1 presents the mean and standard deviation of concentrations (ppm) for all reported elements from each site. As discussed above, the first objective of this project was to test the ability of EDXRF toward the differentiation of intra-island basalt procurement sites, and it is possible to differentiate between each analyzed site using the EDXRF elemental concentrations reported. Simple biplots of the EDXRF elemental concentration (ppm) data achieve clear separation between the sites while displaying intra-site cohesion of samples with little or no observable inter-site overlap. Figure 7.3 is a biplot of titanium (Ti) and magnesium (Mg) concentrations that displays separation between all sites, with samples from the Pago volcanic province sites of Alega and Asiapa displaying the least amount of internal cohesion and some overlap with the Taputapu site of Tataga-matau. Figure 7.4 displays differentiation of all four sites through a biplot of Ti and calcium (Ca), while again samples from Alega and Asiapa display the least amount of intra-site cohesion and some overlap with Tataga-matau. Although these biplots of EDXRF elemental concentrations display some overlap between several samples from the two Pago volcanic sites, it is important to note that overall there is clear differentiation across and within intra-island volcanic provinces.

The second objective of this project was to compare the results of EDXRF analysis against the previous application of INAA (Johnson et al. 2007) on the same samples from the same sites. The purpose of this comparison is to investigate the efficacy of EDXRF for the differentiation of intra-island sites against differentiation achieved using INAA. Elemental concentrations for INAA on the same 72 samples from the sites of Alega, Asiapa, Lau'agae, and Tataga-matau are not presented in this chapter but are reported by Johnson et al. (2007). The discussion of EDXRF data compared against INAA data is presented primarily through comparison of elemental concentration biplots, but also through the comparison of results from exploratory multivariate statistical analyses.

Initially, EDXRF and INAA data were compared through biplots of elemental concentrations (ppm) for Ti, manganese (Mn) and aluminum (Al). Figure 7.5 is a biplot of Ti and Al concentrations from EDXRF that displays intra-site group cohesion and clear separation between the four sites; while the INAA concentrations for Ti and Al (Fig. 7.6) produce a similar trend in differentiation for the same samples but display less evident intra-site cohesion and inter-site separation. A similar relationship between EDXRF and INAA data is evident in biplots for Ti and Mn concentrations. A biplot of EDXRF concentrations for Ti and Mn (Fig. 7.7) displays differentiation between sites, but a biplot for the same elements from INAA concentrations (Fig. 7.8) does not display analogous inter-site differentiation. Although the biplots of concentrations for certain elements reported for both EDXRF and INAA do not display similar levels of intra-site cohesion or inter-site differentiation, there is a linear relationship in the separation of samples and sites that is evident in all the biplots for

Table 7.1 Mean and standard deviation for 15 analytes reported from EDXRF on the four fine-grained basalt procurement sites from Tutuila, American Samoa

Element	Lau'agae (n = 18)	Asiapa (n = 18)	Tataga-matau (n = 18)	Alega (n = 18)
Al	89349.16 ± 1860.93	89759.33 ± 7715.37	86476.51 ± 968.37	85807.30 ± 2610.57
Ca	11560.67 ± 311.75	13665.86 ± 467.48	11990.28 ± 337.30	12899.94 ± 395.70
Cu	25.00 ± 7.07	10.00 ± 0.00	14.44 ± 6.16	11.11 ± 3.23
Fe	100531.11 ± 1479.17	90697.78 ± 5238.21	97585.00 ± 2084.18	97667.78 ± 2106.797
K	13428.29 ± 362.12	15873.58 ± 542.10	13927.32 ± 391.79	14983.93 ± 459.62
Mg	25789.56 ± 507.82	21769.23 ± 1141.80	24806.84 ± 441.63	22750.61 ± 641.77
Mn	1243.87 ± 42.59	1409.52 ± 98.50	1351.86 ± 52.07	1324.76 ± 58.23
Rb	31.67 ± 3.84	40.00 ± 0.00	32.78 ± 4.61	37.78 ± 4.28
Si	232030.46 ± 1865.012	243541.95 ± 2084.69	236857.56 ± 2008.36	241356.93 ± 1464.98
Sr	791.67 ± 15.811	743.89 ± 54.44	720.00 ± 18.15	718.89 ± 19.37
Ti	24219.54 ± 400.93	18163.82 ± 952.69	21566.37 ± 508.73	19465.09 ± 463.53
V	302.78 ± 10.18	251.67 ± 12.95	290.56 ± 14.34	286.11 ± 15.39
Y	41.11 ± 4.71	56.67 ± 8.402	42.22 ± 4.28	51.11 ± 3.23
Zn	170.56 ± 8.02	182.78 ± 14.061	167.78 ± 6.47	176.11 ± 11.95
Zr	89349.16 ± 1860.93	463.33 ± 21.69	401.67 ± 15.44	478.89 ± 11.32

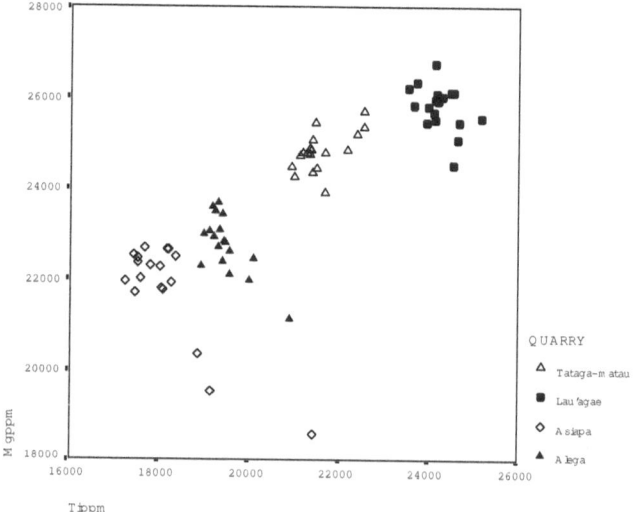

Fig. 7.3 Biplot of Ti and Mg concentrations (ppm) from energy dispersive X-ray fluorescence (EDXRF) data

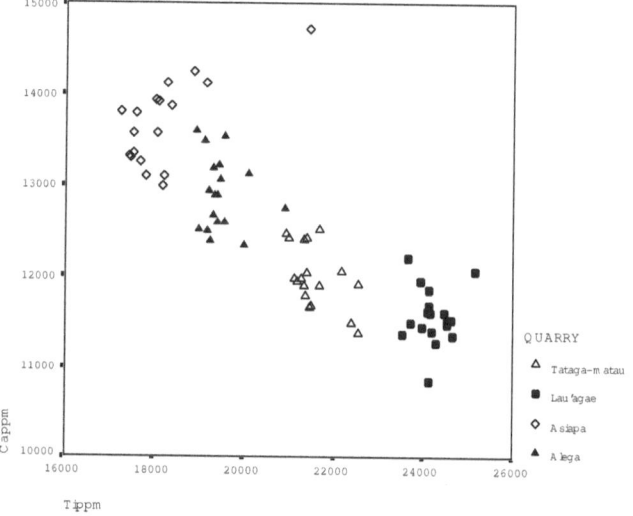

Fig. 7.4 Biplot of Ti and Ca concentrations (ppm) from EDXRF data

both the EDXRF and INAA concentrations. In a previous application of XRF on basalt samples from Tutuila, Clark et al. (1997, p. 75) note a similar trend in the differentiation of multiple intra-island samples and sites and remark that, "Although the quarry samples fall into fairly well-defined groups that define a single fractionation trend on all applicable plots of major and trace elements, there is considerable overlap in quarries, even some that are widely separated geographically."

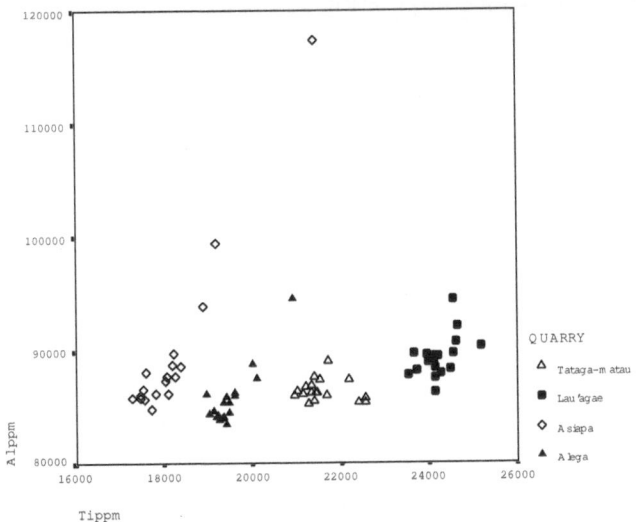

Fig. 7.5 Biplot of Ti and Al concentrations (ppm) from EDXRF data

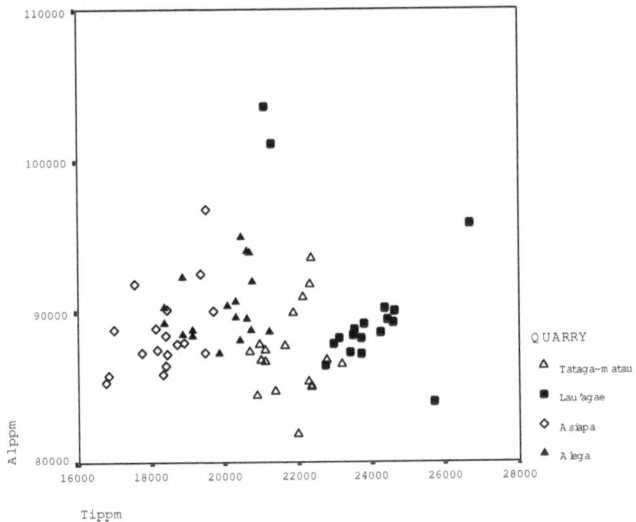

Fig. 7.6 Biplot of Ti and Al concentrations (ppm) from instrumental neutron activation analysis (INAA) data

As the final step in this investigation of the EDXRF data, multivariate statistical analyses were applied to further explore variability and test the group cohesion between each site. Multivariate statistical analyses were also used in an attempt to mitigate the linear trend and overlap in site differentiation through the inclusion of multiple variables to define group cohesion and separation. EDXRF concentrations reported in table 7.1 as well as INAA concentrations for those 15 elements derived

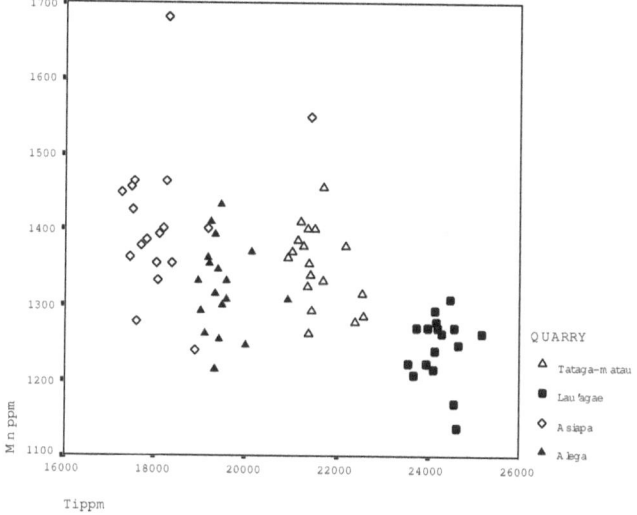

Fig. 7.7 Biplot of Ti and Mn concentrations (ppm) from EDXRF data

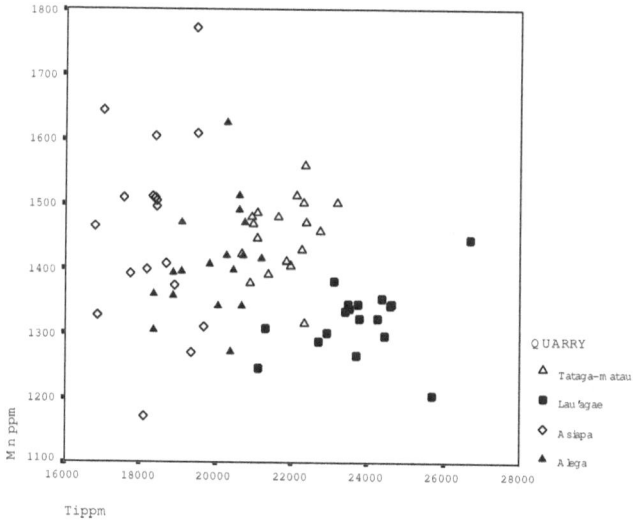

Fig. 7.8 Biplot of Ti and Mn concentrations (ppm) from INAA data

from the same 72 samples (Johnson et al. 2007) were included in the multivariate statistical analysis. Principal component analysis (PCA) was used to further explore and classify possible groups beyond bivariate relationships, and then canonical discriminant analysis (CDA) was applied to confirm both bivariate and multivariate group affiliations (Glascock et al. 1998; see also Chap. 8). All data were Log (base

10) transformed (Baxter 1994) prior to multivariate exploratory statistical analysis, and all multivariate statistical methods were conducted with SPSS version 11 for Mac OSX.

When the elemental concentration data for the 15 elements reported for EDXRF was explored using PCA, the first two principal component scores represented over 71% of the total variability for the dataset, while the first two PCA scores of the INAA concentration data represented 67% of the variability for the same set of samples. Biplots of the first two PCA scores were produced for both the EDXRF (Fig. 7.9) and INAA (Fig. 7.10) elemental concentration data. The biplots of PCA scores for EDXRF and INAA data display dissimilar levels of inter-site differentiation and intra-site cohesion of samples as evident in the elemental concentration biplots and do not appear to display any subgrouping. The biplot of PCA scores for EDXRF data again displays a high level of intra-group cohesion and clear differentiation between the sites. The PCA data for INAA concentrations displays a linear trend in intra-site clustering of samples and less evident differentiation between groups. Although the majority of PCA data for EDXRF clustered tightly, the linear trend more evident in the INAA plot is once again apparent in the samples from Alega and Asiapa. After groups were classified through bivariate and multivariate analyses, a stepwise CDA was used to confirm the apparent group affiliation of the 18 samples for each of the assigned basalt procurement sites. For both the EDXRF and INAA datasets, all 72 samples were assigned to the proper procurement site (or group) with no less than 95% confidence, and at least 95% of the total variability for both datasets was represented in the first two discriminant functions. Biplots of the first two discriminant functions for both the EDXRF (Fig. 7.11) and INAA (Fig. 7.12) datasets are presented to display the differentiation of procurement sites as confirmed by canonical discriminant analyses. The plot of CDA functions for EDXRF concentrations again displays high intra-group cohesion and shows very discrete separation of each site, and displays no overlap of the Pago Volcanic sites of Alega and Asiapa. The plot of CDA functions for the INAA data also displays differentiation of each group including the previously overlapped Alega and Asiapa groups, but displays some overlap between the geographically isolated Lau'agae and Tataga-matau groups that is not evident in the same EDXRF plot. After applying multivariate exploratory analyses to the EDXRF concentration data, both PCA and CDA confirm the expected intra-group cohesion and inter-group separation that was initially determined through bivariate plots of elemental concentrations.

Conclusions and Final Discussion

This research has achieved the two primary objectives set forth earlier in the chapter. The first objective was the differentiation of multiple intra-island basalt procurement and manufacture sites on Tutuila through EDXRF elemental compositional data. The four clearly defined groups displayed through exploratory analysis are consistent with the expectation that the 18 samples from each individual site

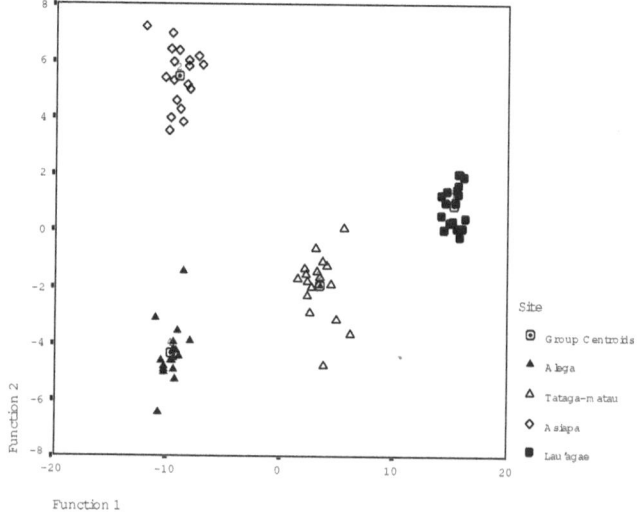

Fig. 7.9 Biplot of first two principal component analysis (PCA) scores from EDXRF data

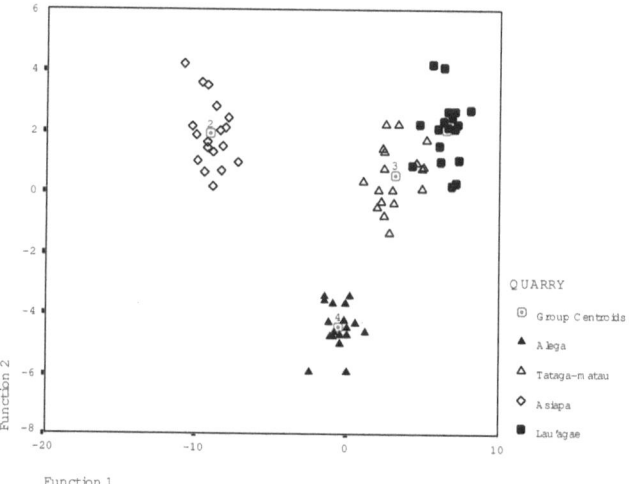

Fig. 7.10 Biplot of first two PCA scores from INAA data

should display group cohesion and that inter-site variability should exceed intra-site variability. The bivariate and multivariate exploratory analyses of EDXRF data display clear separation of each individual site while maintaining a high level of internal cohesion with little or no inter-site overlap between the 72 samples. These classification results were then further supported through discrimination using stepwise CDA, which confirmed the unambiguous differentiation of each individual fine-grained basalt procurement site.

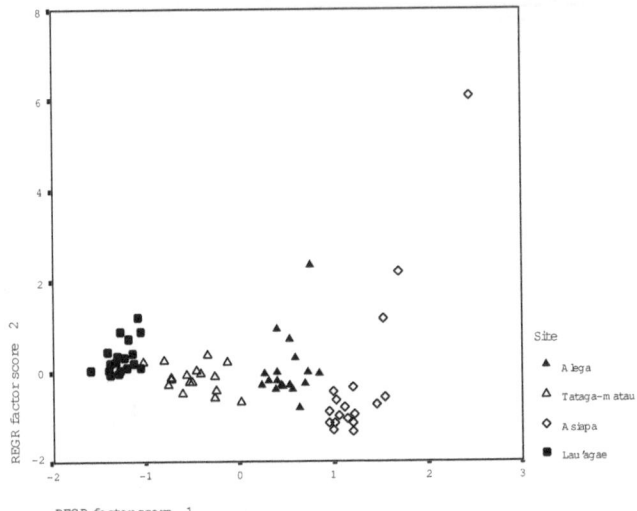

Fig. 7.11 Biplot of first two canonical discriminant analysis (CDA) functions from EDXRF data

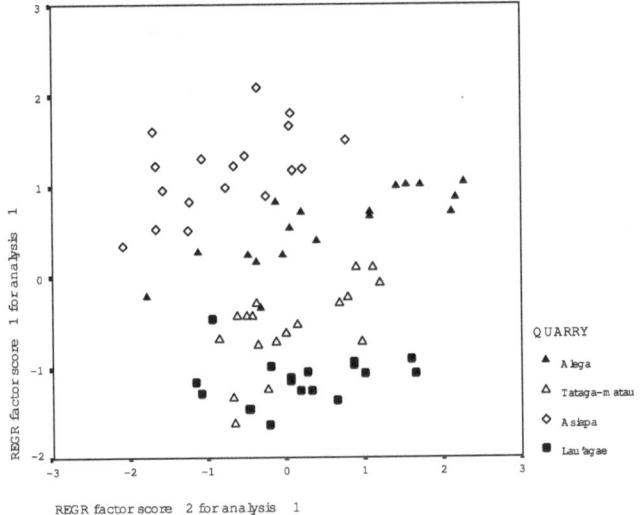

Fig. 7.12 Biplot of first two CDA functions from INAA data

The second objective of this research was to assess the efficacy of EDXRF in comparison to previous applications of INAA on the same samples. This chapter was not intended as a discussion of the analytical precision or capabilities of EDXRF or INAA. It was designed to discuss the suitability of EDXRF and INAA toward the differentiation of these specific sites and samples, and the implications therein for future provenance analysis of fine-grained basalts from Tutuila. As discussed in the previous section, both EDXRF and INAA compositional data can be utilized for the differentiation of the expected groups, but the EDXRF data appears to display greater intra-site cohesion and inter-site separation for this particular set of samples. When compared with INAA, the EDXRF compositional data provides a similar or higher level of differentiation between sites achieved through both biplots of compositional variability as well as exploratory multivariate statistical analyses. These preliminary comparative results have led the author to the determination that EDXRF is an appropriate technique for the analysis of fine-grained basalt procurement and tool production on the island of Tutuila.

The results of this comparison of EDXRF and INAA are not to be interpreted as a commentary on the analytical superiority or inferiority of either technique discussed. The dilemma for archaeologists attempting to determine which instrument of elemental analysis is "the best choice" for archaeological applications has been repeatedly discussed in the archaeometry literature (Bishop et al. 1990; Neff 2000; Shackley 1998a). The editor of this volume, Shackley, addressed this very question by stating, "It depends...the problem of design and the level of precision needed to address that design will determine which instrument is the best for a given project (Shackley 1998b, p. 7)." Keeping in mind the relative nature of "best technique" as described by Shackley, and considering the dominance of EDXRF over INAA as a technique of choice toward Polynesia basalt provenance studies, the results of this research suggest that EDXRF is currently a more suitable technique than INAA for the analysis of Tutuila fine-grained basalt sources. That endorsement must be tempered with the final caveat that this analysis is preliminary and includes a very limited sample of the multiple procurement and production sites on Tutuila. As more sites are sampled and analyzed, the results reported in this chapter may no longer be applicable. As project designs are adapted for changing research areas and project goals, no single method of analysis may provide a clear characterization and differentiation of all sites, and it is necessary to continue the evaluation of multiple instruments and methodologies to ensure that the best technique or combination of techniques for a particular research design is chosen to address the future investigation of basalt sources and artifacts on the island of Tutuila, in Polynesia and around the globe.

Acknowledgments Field sampling and research for this project were conducted partially through funding from the American Samoa Historic Preservation Office and the Texas A&M Office of the Vice President for Research. I would first like to thank Dr. William D. James, Director of the EAL at Texas A&M University, for his advice and support and for allowing access to the facilities of the EAL without which this research would not have been possible. I am deeply grateful for the guidance and encouragement of Dr. Peter Mills and Dr. Steve Lunblad from the University of Hawaii at Hilo who shared their EDXRF method and experience analyzing Polynesia basalt artifacts and sources, which lead directly to the calibration of the EDXRF method employed for this project at the Texas A&M EAL. Also, the advice and assistance of Dr. Suzanne L. Eckert was instrumental in the inception and completion of this project. Finally, I need to thank Michael Raulerson, David Foxe, and Daniel Welch for their assistance in the preparation of samples for EDXRF analysis. As always *fa'afetai tele* to the wonderful people of Tutuila.

References

Allen, M.S., & Johnson, K.T.M. (1997) Tracking ancient patterns of interaction: recent geochemical studies in the southern cook islands. In M.I. Weisler (Ed.), *Prehistoric long-distance interaction in Oceania: an interdisciplinary approach* (pp. 111–113). Auckland: New Zealand Archaeological Association.

Baxter, M.J. (1994) *Exploratory multivariate analysis in archaeology*. Edinburgh: Edinburgh University Press.

Best, S., Leach, H.M., & Witter, D.C. (1989) *Report on the second phase of fieldwork at the Tataga-matau site, American Samoa, July–August 1988*. Dunedin, NZ: Department of Anthropology, University of Otago.

Best, S., Sheppard, P.J., Green, R.C., & Parker, R. (1992) Necromancing the stone: archaeologists and adzes in Samoa. *Journal of the Polynesian Society*, 101, 45–85.

Bishop, R.L., Canouts, V., Crown, P.L., & De Atley, S.P. (1990) Sensitivity, precision, and accuracy: their roles in ceramic compositional data bases. *American Antiquity*, 55, 537–546.

Buck, P.H. (1930) *Samoan material culture*. Honolulu, HI: Bernice P. Bishop Museum Press.

Clark, J. T. .(1980) Historic Preservation in American Samoa: Program Evaluation and Archaeological Site Inventory. Unpublished manuscript, Bernice P. Bishop Museum, Honolulu HI.

Clark, J.T. (1989) *The Eastern Tutuila Archaeological Project, 1988. Final report*. Unpublished Report for the Government of America Samoa Office of Historic Preservation, Fargo, ND: North Dakota State University.

Clark, J.T. (1992) *The archaeology of Alega valley – residence and small industry in prehistoric Samoa*. Unpublished Report for the Government of America Samoa Office of Historic Preservation, Fargo, ND: North Dakota State University.

Clark, J.T., Wright, E., & Herdrich, D.J. (1997) Interactions within and beyond the Samoan archipelago: evidence from basaltic rock geochemistry. In M.I. Weisler (Ed.), *Prehistoric long-distance interaction in Oceania: an interdisciplinary approach* (pp. 68–83). Auckland: New Zealand Archaeological Association.

Collerson, K.D., & Weisler, M.I. (2007) Stone adze compositions and the extent of ancient Polynesian voyaging and trade. *Science*, 317, 1907–1911.

Crews, C.T. (2008) *The lithics of Aganoa village (AS-22-43), American Samoa: a test of chemical characterization and sourcing Tutuilan tool-stone*. Unpublished M.A. thesis, College Station, TX: Texas A&M University.

Davidson, J.M. (1977) Western Polynesia and Fiji: prehistoric contact, diffusion and differentiation in adjacent archipelagos. *World Archaeology*, 9, 82–94.

Glascock, M.D. (1992) Neutron activation analysis. In H. Neff (Ed.), *Chemical characterization of ceramic pastes in archaeology* (pp.11–26). Madison, WI: Monographs in World Archaeology No. 7. Prehistory Press.

Glascock, M.D., Braswell, G.E., & Cobean, R.H. (1998) A systematic approach to obsidian source characterization. In M.S. Shackley (Ed.), *Archaeological obsidian studies: method and theory* (pp. 15–66). New York: Plenum Press.

Green, R.C. (1962) Obsidian, its application to archaeology. *New Zealand Archaeological Association Newsletter*, 5, 8–16.

Green, R.C. (1964) Sources, ages and exploration of New Zealand obsidian: an interim report. *New Zealand Archaeological Association Newsletter*, 7, 134–143.

Green, R.C. (1998) A 1990's perspective on method and theory in archaeological volcanic glass studies. In M.S. Shackley (Ed.), *Archaeological obsidian studies: method and theory* (pp. 223–235). New York: Plenum Press.

Heath T. (1840) The Navigator's or Samoan Islands. Their manners, customs and superstitions. The Polynesian 19 February, Vol. I, No. 15, Honolulu.

James, W.D., Raulerson, M., & Johnson, P.R. (2007) Archaeometry at Texas A&M University: a characterization of Samoan basalts. *Archaeometry*, 49, 395–403.

Johnson, P.R. (2005) *Instrumental neutron activation analysis (INAA) characterization of pre-contact basalt quarries on the American Samoan Island of Tutuila.* Unpublished M.A. thesis, College Station, TX: Texas A&M University.

Johnson, P.R., Pearl, F.B., Eckert, S.L., & James, W.D. (2007) INAA of pre-contact basalt quarries on the Samoan Island of Tutuila: a preliminary baseline for an artifact-centered provenance study. *Journal of Archaeological Science, 34*(7), 1078–1087.

Kaeppler, A.L. (1978) Exchange patterns in goods and spouses: Fiji, Tonga and Samoa. *Mankind,* 11, 246–252.

Kahn, J.G. (2005) *Household and community organization in the Late Prehistoric Society Islands (French Polynesia).* Unpublished Ph.D. dissertation, Berkeley: University of California-Berkeley.

Kikuchi, W.K. (1963) *Archaeological surface ruins in American Samoa.* Unpublished M.A. thesis, Honolulu, HI: University of Hawai'i.

Kirch, P.V., & Green, R.C. (2001). *Hawaiki, ancestral Polynesia: an essay in historical anthropology.* Cambridge: Cambridge University Press.

Leach, H.M., & Witter, D.C. (1985) *Final project report on the survey of the Tataga-Matau fortified quarry complex, near Leone, American Samoa.* Dunedin, NZ: University of Otago.

Leach, H.M., & Witter, D.C. (1987). Tataga-matau "rediscovered." *New Zealand Journal of Archaeology, 9,* 33–54.

Leach, H.M., & Witter, D.C. (1990) Further investigations at the Tataga-matau site, American Samoa. *New Zealand Journal of Archaeology, 12,* 51–83.

Lebo, S.A., & Johnson, K.T.M. (2007) Geochemical sourcing of rock specimens and stone artifacts from Nihoa and Necker Islands, Hawai'i. *Journal of Archaeological Science, 34,* 858–871.

Lunblad, S.P., Mills, P.R., & Hon, K. (2008) Analysing archaeological basalt using non-destructive x-ray fluorescence (EDXRF): effects of post-depositional chemical weathering and sample size on analytical precision. *Archaeometry, 50,* 1–11.

MacDonald, G.A. (1944) Petrography of the Samoan Islands. *Bulletin of the Geological Society of America, 55,* 1333–1362.

MacDonald, G.A. (1968) Contribution to the petrology of Tutuila, American Samoa. *Geologische Rundschau, 57,* 821–837.

MacDougall, I. (1985) Age and evolution of the volcanoes of Tutuila, American Samoa. *Pacific Science, 39,* 311–320.

Mills, P.R., Lunblad, S.P., Smith, J.G., McCoy, P.C., & Naleimaile, S.P. (2008) Science and sensitivity: a geochemical characterization of the Mauna Kea adze quarry complex, Hawai'i Island, Hawaii. *American Antiquity, 73,* 743–758.

Moore, J.R., Kennedy, J. (1996). Archaeological Resources on Lau'agae Ridge: A Phase II Cultural Resource Evaluation of Site AS-21-100 (The Lau'agae Ridge Quarry) for Phase III of the Onenoa Road Project Located in East Vaifanua County, Tutuila Island, American Samoa March 1996. Archaeological Consultants of Hawaii, Inc., Pago.

Natland, J.H. (1980) The progression of volcanism in the Samoan linear volcanic chain. *American Journal of Science, 280,* 709–735.

Natland, J. (2003) The Samoan Chain: A Shallow Lithospheric Fracture System, http://www.mantleplumes.org/Samoa.html

Neff, H. (2000) Neutron activation analysis for provenance determination in archaeology. In J.D. Winefordner (Ed.), *Chemical analysis: a series of monographs on analytical chemistry and its applications* (pp. 81–127). New York: Wiley.

Parker, R., & Sheppard, P.J. (1997) Pacific island adze geochemistry studies at the University of Auckland. In M.I. Weisler (Ed.), *Prehistoric long-distance interaction in Oceania: an interdisciplinary approach* (pp. 205–211). Auckland: New Zealand Archaeological Association.

Rolett, B.V., Conte, E., Pearthree, E.J., & Sinton M. (1997) Marquesan voyaging: archaeometric evidence for inter-island contact. In M.I. Weisler (Ed.), *Prehistoric long-distance interaction in Oceania: an interdisciplinary approach* (pp. 134–148). Auckland: New Zealand Archaeological Association.

Shackley, M.S. (1998a). Gamma rays, x-rays and stone tools: some recent advances in archaeo-
    logical geochemistry. *Journal of Archaeological Science*, 25, 259–270.
Shackley, M.S. (1998b) Current issues and future directions in archaeological volcanic glass
    studies: an introduction. In M.S. Shackley (Ed.), *Archaeological obsidian studies: method
    and theory* (pp. 1–12). New York: Plenum Press.
Sheppard, P.J., Walter, R., & Parker, R. (1997) Basalt sourcing and the development of Cook
    Island exchange systems. In M.I. Weisler (Ed.), *Prehistoric long-distance interaction in
    Oceania: an interdisciplinary approach* (pp. 205–211). Auckland: New Zealand Archaeolog-
    ical Association.
Stearns, H.T. (1944) Geology of the Samoan Islands. *Bulletin of the Geological Society of
    America*, 55, 1279–1332.
Weisler, M.I. (1990) A technological, petrographic, and geochemical analysis of the Kapohaku
    adze quarry, Lana'i, Hawaiian Islands. *New Zealand Journal of Archaeology*, 12, 29–50.
Weisler, M.I. (1993a) Chemical characterization and provenance of Manu'a adz material using
    non-destructive x-ray fluorescence technique. In P.V. Kirch, T.L. Hunt (Eds.), *The To'aga site:
    three millennia of Polynesian occupation in the Manu'a islands, American Samoa*
    (pp. 167–187). Berkeley: University of California Archaeological Research Facility.
Weisler, M.I. (1993b) Provenance studies of Polynesian basalt adze material: a review and
    suggestions for improving regional databases. *Asian Perspectives*, 32, 61–83.
Weisler, M.I. (1997) Prehistoric long distance interaction at the margins of Polynesia. In
    M.I. Weisler (Ed.), *Prehistoric long-distance interaction in Oceania: an interdisciplinary
    approach* (pp. 149–172). Auckland: New Zealand Archaeological Association.
Weisler, M.I. (1998) Hard evidence for prehistoric interaction in Polynesia. *Current Anthropology*,
    39, 521–532.
Weisler, M.I. (2002) Centrality and collapse of long-distance voyaging in East Polynesia. In
    M. Glascock (Ed.), *Geochemical evidence for long distance exchange* (pp. 257–273).Westport,
    CT: Bergin and Garvey.
Weisler, M.I. (2003) A stone tool basalt source on 'Ata southern Tonga. *New Zealand Journal of
    Archaeology*, 25, 113–120.
Weisler, M.I., & Kirch, P.V. (1996) Interisland and interarchipelago transfer of stone tools in
    prehistoric Polynesia. *Proceedings of the National Academy of Sciences USA*, 93, 138–185.
Weisler, M.I., & Sinton, J.M. (1997) Towards identifying prehistoric interaction systems in
    Polynesia. In M.I. Weisler (Ed.), *Prehistoric long-distance interaction in Oceania: an inter-
    disciplinary approach* (pp. 173–193). Auckland: New Zealand Archaeological Association.
Weisler, M.I., & Woodhead, J.D. (1995) Basalt Pb isotope analysis and the prehistoric settlement
    of Polynesia. *Proceedings of the National Academy of Sciences U S A*, 92, 1881–1885.
Weisler, M.I., Kirch, P.V., & Endicott, J.M. (1994) The Mata'are basalt source: implications for
    prehistoric interaction studies in the Cook Islands. *Journal of the Polynesian Society*, 103,
    203–216.
Winterhoff, E.Q. (2007) *The political economy of ancient Samoa: basalt adze production and
    linkages to the social status*. Unpublished Ph.D. dissertation, Eugene: University of Oregon.
Winterhoff, E.Q., Wozniak, J.A., Ayres, W.S., & Lash, E. (2007) Intra-island source variability on
    Tutuila, American Samoa and prehistoric basalt adze exchange in Western Polynesia-Island
    Melanesia. *Archaeology in Oceania*, 42(2), 65–71.
Wright, Elizabeth (1986). Petrology and geochemistry of shield-building and post-erosional lava
    series of samoa: implications for mantle heterogeneity and magma genesis (Ocean Island,
    Volcano, Ultramafic xenolith). Ph.D. dissertation, University of California, San Diego, United
    States – California.

Chapter 8
Comparison and Contrast Between XRF and NAA: Used for Characterization Of Obsidian Sources in Central Mexico

Michael D. Glascock

Introduction

Since the mid-1950s, the use of physical and chemical techniques to examine objects of archaeological and historical importance has grown rapidly (Glascock 2008; Guerra 2008). Information gathered from the examination of artifacts, paintings, sculptures, and other materials is frequently used to: (1) answer questions about authenticity; (2) assign artifacts to particular time periods; (3) determine the provenance of artifacts; (4) investigate the technologies used to manufacture the objects under examination; and (5) study changes in museum objects induced by the effects of aging.

It is desirable that the techniques employed when examining artifacts and museum objects be (1) rapid, such that large numbers of samples can be studied; (2) versatile, to allow objects of various types, shapes, and sizes; and (3) sensitive to large numbers of major, minor, and trace elements. The analytical data should be accurate and reproducible such that the data generated by different techniques and from different laboratories are in agreement. Interpretation of the analytical data with regard to the groupings of source samples and about the assignment of artifacts to these groups should also agree. Analytical techniques that are nondestructive or those require minimal sampling are preferred over destructive techniques that require removal of sample material for dissolution, etc.

There is no single analytical technique capable of answering every possible archaeological question concerning composition. However, among the modern analytical techniques that satisfy most of the demands of archaeologists, museum curators, and others are: X-ray fluorescence spectrometry (XRF), neutron activation analysis (NAA), and laser ablation-inductively coupled plasma-mass spectrometry (LA-ICP-MS). All three analytical techniques have been making significant

M.D. Glascock (✉)
University of Missouri, Research Reactor Center, Columbia, MO 65211, USA
e-mail: glascockm@missouri.edu

M.S. Shackley (ed.), *X-Ray Fluorescence Spectrometry (XRF) in Geoarchaeology*,
DOI 10.1007/978-1-4419-6886-9_8, © Springer Science+Business Media, LLC 2011

contributions to modern archaeological and museum research because they can be performed instrumentally and without sample dissolution. Measurements on artifacts and art objects can be performed in a completely nondestructive manner using energy dispersive X-ray fluorescence (EDXRF), but the sensitivity is limited for some elements. Measurements by LA-ICP-MS can be performed on sample surfaces with minimal destruction, but standardization is difficult and the sample must fit inside the laser ablation chamber. High-precision analyses by NAA can be designed in such a way that minimal damage is performed on the object by drilling from an obscure location that does not harm the overall appearance of the object or by using smaller samples than the other techniques, but NAA is more expensive, time consuming, and less available.

The purpose of this chapter is to: (1) describe and compare the similarities and differences between NAA and XRF; (2) describe a calibration procedure employed by the Archaeometry Lab at the University of Missouri Research Reactor (MURR) that facilitates collection of comparable data on archaeological obsidian by both techniques; (3) present a comprehensive geochemical database on obsidian sources in central Mexico created using both NAA and XRF; and (4) recommend a procedure for using the information in this database to determine the provenance of obsidian artifacts with a high-level of confidence. This is important information for archaeologists working in central Mexico because obsidian is one of the most abundant and popular archaeological materials. With this knowledge, archaeologists can reduce labor, analysis times, and analytical costs when analyzing obsidian artifacts from central Mexico.

Description of XRF

The XRF technique has been described adequately in other chapters of this book and extensively elsewhere (Jenkins 1999). Therefore, the description presented here will be considerably abbreviated.

X-ray fluorescence is a two-step process (see also Chaps. 2 and 3). The first step involves an X-ray photon produced by an X-ray tube or a low-energy photon emission source which strikes an atom in the sample and creates a vacancy by knocking out an inner-shell electron. The second step is a readjustment of the atom which occurs by filling the inner-shell vacancy with an outer-shell electron and the simultaneous emission of a new X-ray photon commonly referred to as a fluorescent X-ray. The first step consumes all of the incident X-ray energy which must be greater than the binding energy of the inner-shell electron. Any excess energy is carried away by the electron in the form of kinetic energy. In the second step, the energy of the fluorescent X-ray corresponds exactly to the difference in energy between the two atomic energy levels and is unique for each element.

The diagram shown in Fig. 8.1 illustrates the process of XRF. The fluorescent X-ray is referred to as a K_α or K_β X-ray if the K-shell electron is replaced by an electron

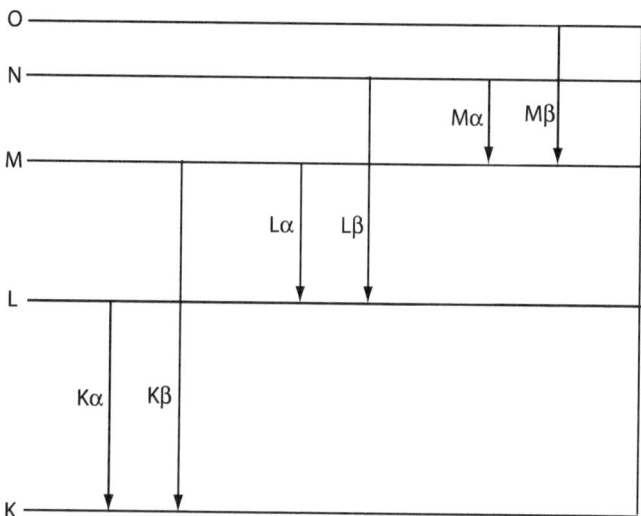

Fig. 8.1 Electron shells from XRF

originally from the L-shell or M-shell, respectively. If an L-shell electron is replaced by an electron from either the M-shell or the N-shell, the fluorescent X-ray is referred to as L_α or L_β X-ray, respectively. Atoms with different numbers of protons (e.g., Fe, Rb, Pb) have different fluorescent X-ray energies as defined by Moseley's Law in which the X-ray energies for each element are proportional to $(Z-1)^2$. The energies of X-rays for an element are the same irrespective of the specific isotope of that element involved in the interaction. Listings of X-ray energies and their fluorescence yields are available (Sansonetti et al. 2005). The energies of X-rays useful for measurement in most archaeological samples are below 40 keV.

When using XRF to analyze a sample, quantification is possible by measuring the intensities of the X-rays observed in the spectrum collected by a multichannel analyzer system. Although in principal, XRF can be used to measure most of the elements in the periodic table, the X-rays for elements below $Z = 11$ are very difficult to measure and special equipment setups are necessary to measure their intensities accurately. For some of the higher Z elements, the L_α and L_β X-rays may be easier to measure than the higher energy K_α and K_β X-rays.

Figure 8.2 shows a typical XRF spectrum measured on a sample of obsidian exposed to a source of X-rays. The intensities of the fluorescent X-rays observed are proportional to the incident flux of X-ray photons, to the concentrations of the elements in the sample, to the efficiency of the detector, and to geometric effects. By comparing the fluorescent X-ray intensities measured from the sample with those obtained from standards, it is possible to calculate concentrations of elements in the sample. However, the accuracies of elemental concentrations determined by XRF are affected by a many factors including surface texture, sample thickness relative to kiloelectron volt energies, inhomogeneities within the sample, particle size, and matrix effects.

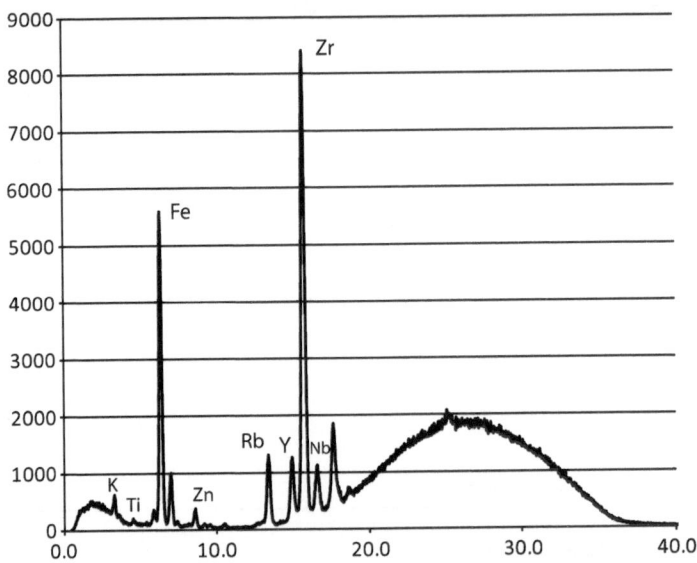

Fig. 8.2 XRF spectrum from sample of obsidian

Matrix effects are caused by the absorption of fluorescent X-rays by coexisting elements in the sample which result in reduced intensity and/or enhancement of fluorescence radiation due to secondary radiation emitted by the fluorescing element or a coexisting element which yields an increase in intensity. The matrix effects observed in XRF are sometimes referred to as mass absorption effects. In general, every element exerts a mass absorption effect on all other elements present in the sample, but some of the elements are more affected than others due to absorption edges. When combined, these effects result in curved rather than linear calibration lines for each element. By using the intensities of the primary X-rays scattered by the sample which are proportional to the effective mass absorption coefficient, corrections to the absorption/enhancement effects can be calculated resulting in a more accurate quantitative evaluation of sample composition. However, it is essential that the samples (i.e., standards) used for XRF calibration be as similar to the unknowns as possible to properly correct these matrix effects.

Over the past few decades, with the development of semiconductor detectors, computerized multichannel analyzers (MCA), and miniaturization of X-ray tubes and other components, significant advances in the field of XRF spectrometry have occurred. When the X-ray photons enter a semiconductor material, they interact primarily by photoelectric absorption to produce electron-hole pairs. The number of pairs produced is proportional to the energy of the incident photon. By applying a voltage across the semiconductor, an electrical signal is created. The signals from different energy photons are processed by the MCA to establish a spectrum of X-ray energies measured by the detector. Prior to the development of the thermo-electric (i. e., Peltier) cooler, semiconductor detectors required a nearby reservoir of liquid nitrogen to maintain the semiconductor at a cold temperature. Miniaturization of components has led to the development of light-weight, battery-powered (or solar cell-powered) XRF spectrometers capable of being operated in almost any environment, including missions on the planet Mars. Many of the newest and most popular selling XRF spectrometers are hand-held units that can be easily carried in a backpack (see Chap. 6).

Description of NAA

NAA differs from XRF because NAA is based on the nuclear properties of the elements in the analytical sample. Many of these properties can be found on the *Chart of the Nuclides* (Baum et al. 2002) which organizes the isotopes by their numbers of neutrons and protons. Some of the most important properties listed for each isotope include the abundances of stable isotopes, half-lives of radioactive isotopes, neutron capture cross sections, and energies of main gamma rays. An even more comprehensive source of information concerning the isotopes is the *Table of the Isotopes* (Firestone et al. 1996).

Nuclear Reactions

As illustrated in Fig. 8.3, when a sample is exposed to thermal neutrons (energies of about ~0.025 eV), usually from a nuclear reactor, some of the atomic nuclei present will undergo nonelastic collisions during which incident neutrons are captured (or absorbed) by the target nucleus. The process of capturing a neutron creates a compound nucleus (X^*) with one additional mass unit. The compound nucleus also has an excess energy of about 8 MeV due to the binding energy of the neutron. In order to rid itself of this excess energy, the compound nucleus undergoes

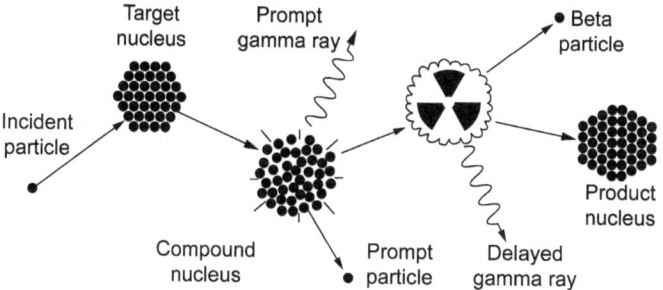

Fig. 8.3 Neutron capture reaction

instantaneous emission of one or more prompt gamma rays (γ_p) until it arrives at its ground state.

In general terms, the neutron capture reaction can be written as:

$$\, _{0}^{1}n_{th} + \, _{Z}^{A}X \xrightarrow{\text{yields}} \, _{Z}^{A+1}X^* \xrightarrow{\text{yields}} \, _{Z}^{A+1}X + \gamma_p, \tag{8.1}$$

or using a more abbreviated notation:

$$A(n, \gamma)B, \tag{8.2}$$

Where n is an incident thermal neutron ($_{0}^{1}n_{th}$), A is the target nucleus ($_{Z}^{A}X$), B is the product nucleus ($_{Z}^{A+1}X$), and γ represents the exiting prompt particle(s).

For each variety of isotope in the sample, the rate at which neutron capture occurs depends on the number of target nuclei present, the incident thermal neutron flux, and the reaction cross section (or probability) for thermal neutron capture. An equation describing the reaction rate for thermal neutrons is:

$$R = n\sigma_{th}\varphi_{th}, \tag{8.3}$$

where n is the number of nuclei of a particular isotope species present, σ_{th} is the thermal neutron capture cross section, and φ_{th} is the thermal neutron flux. Neutrons with higher energies ranging up to about 100 keV are referred to as epithermal neutrons and they can also induce (n,γ) reactions. The reactions induced by epithermal neutrons add an additional term to the reaction rate equation as follows:

$$R = n[\sigma\varphi_{th} + I_0\varphi_{epi}], \tag{8.4}$$

where I_0 and φ_{epi} are defined as the epithermal neutron cross section (also known as the resonance integral) and epithermal neutron flux, respectively.

Samples for NAA should not contain major amounts of elements with large neutron capture cross sections (e.g., boron, cadmium, gadolinium, gold, silver) because a significant fraction of the incident neutron flux will be absorbed – causing the need for complicated corrections. Fortunately, the vast majority of archaeological and geological samples for NAA have as their main constituents the elements Si, Al, Mg, Na, K, Ca, Ti, and Fe all of which have very low (n,γ) cross sections. As a result, most samples absorb such a small fraction of the neutron flux that the samples can be considered as uniformly irradiated. For this reason, NAA is considered to be a bulk analysis technique for most geological sample matrices, including obsidian.

Radioactive Decay

The nucleus present in its ground state immediately after the emission of prompt gamma rays can be stable or radioactive. This can be determined by examining the location of the nucleus on the *Chart of the Nuclides*. If the product nucleus is stable, no more gamma rays will be emitted after the sample is removed from the neutron flux. On the other hand, if the nucleus is radioactive, it will have a characteristic rate of decay commonly referred to as the half-life, and emission of one or more delayed gamma rays (γ_d) is likely. Depending upon the particular radioactive isotope created, the half-life can range from milliseconds to several years.

After activation by neutrons, the radioactive nucleus undergoes decay by emitting an electron (β^-) or a positron (β^+). The emission of an electron or positron can be considered equivalent to converting one of the neutrons or protons in the nucleus into a proton or neutron thus conserving charge. Equations describing the β^- and β^+ decay processes, respectively, are as written follows:

$$_{Z}^{A+1}X \xrightarrow{\text{yields}} {}_{Z+1}^{A+1}X + \beta^- + v^-[+\gamma_d], \tag{8.5}$$

$$_{Z}^{A+1}X \xrightarrow{\text{yields}} {}_{Z+1}^{A+1}X + \beta^+ + v[+\gamma_d]. \tag{8.6}$$

The β^- and β^+ decay processes produce particles with a continuum of energies shared between the electron and an antineutrino (v^-) or the positron and a regular neutrino (v), respectively.

Some of the radioisotopes produced by neutron activation are known as "pure β-emitters" in which the parent nucleus decays directly to the ground state of the daughter nucleus by emitting only β particles. These radioisotopes cannot normally be measured by NAA because there are no delayed gamma rays to be measured. However, for the vast majority of radioisotopes, one or more delayed gamma rays will follow β decay. These delayed gamma rays have characteristic energies defined by the differences in the nuclear energy levels (i.e., excited states) of the daughter nucleus in which they occur. For example, Fig. 8.4 shows the decay scheme for the β^- decay of the parent nucleus ^{56}Mn into the excited states of the daughter nucleus ^{56}Fe. The upper excited states are populated by specific percentages of the β decay events, and these excited states depopulate by emission of gamma rays, until the nucleus arrives in the ground state (lowest energy level) condition. Additional radioactive decays are possible if the ground state of the daughter nucleus also undergoes β decay. In addition to unique gamma-ray energies, the gamma-ray branching ratios (i.e., gamma-ray abundances) are also characteristic properties of the radioactive decay sequence.

In NAA, samples and standards are usually irradiated for a known period of time T_i, removed from the neutron flux, and allowed to decay for a time T_d, after which

Fig. 8.4 Decay scheme of
Mn-56 created from neutron
capture

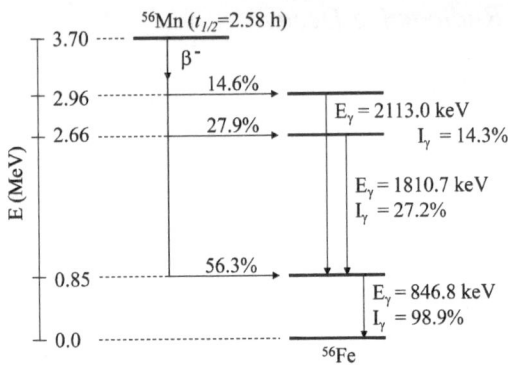

the gamma rays are measured. At any time, the rate of decay for a particular radioisotope depends on the number of radioactive nuclei produced during irradiation, minus the number that decayed during irradiation, and the number that decay during the interval between end of irradiation and beginning of measurement. An equation that expresses the activity of the radioisotope isotope at the beginning of the measurement period is:

$$A = R(1 - e^{-\lambda T_i})e^{-\lambda T_d}, \tag{8.7}$$

where $\lambda = \ln(2)/T_{1/2}$ and $T_{1/2}$ is the half-life of the particular radioisotope. An equation describing the number of radioactive atoms of the isotope of interest that decay between the start and end of a measurement period is:

$$\Delta N = \frac{R}{\lambda}(1 - e^{-\lambda T_i})e^{-\lambda T_d}(1 - e^{-\lambda T_c}). \tag{8.8}$$

For typical NAA measurements, short irradiation and short decay times are used to measure radioisotopes with short half-lives and long irradiation and long decay times are used to measure radioisotopes with long half-lives. When several radioisotopes are produced by NAA such that interferences for gamma rays from different radioisotopes occur, adjustments in irradiation, decay, and counting times can be used to reduce possible interferences and to increase the peak-to-background ratios, etc. Using several measurements, it is possible to measure 30–40 elements in most geological samples by NAA.

Measurement of Gamma Rays

Gamma rays can interact with matter in several ways (i.e., the photoelectric effect, Compton scattering, and pair production), and by each process, they will transfer

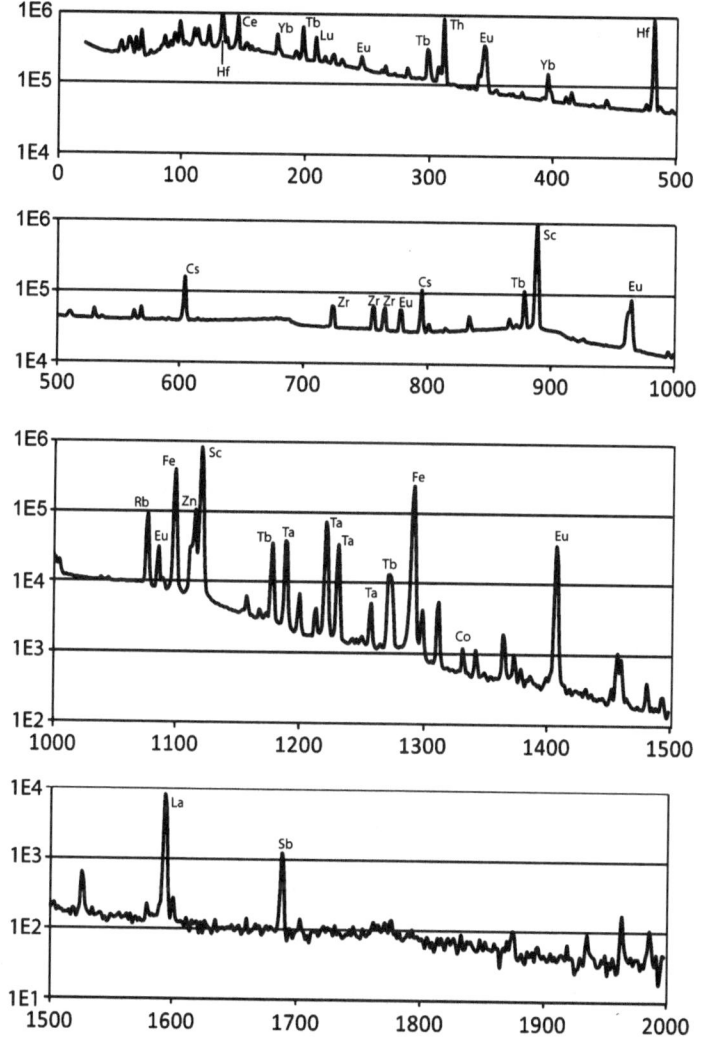

Fig. 8.5 Gamma-ray spectrum for a sample of obsidian

some or all of the incident gamma ray energy into photoelectrons inside a high-purity germanium (HPGe) detector. The photoelectrons are converted into electrical pulses which define the energy of the gamma ray. A gamma-ray spectrum collected on a sample of obsidian after neutron irradiation is shown in Fig. 8.5.

The main gamma rays associated with the radioisotopes present in the sample are labeled on the figure. The actual number of gamma rays observed for each radioisotope is smaller than the total number of decays due to several factors: (1) the abundances of gamma rays are in most cases less than 100% because the gamma ray of a particular energy may is not be released with every decay event; (2) most of the emitted gamma rays do not reach the detector. Due to the isotropic nature of gamma-ray emission, only those headed in the direction of the detector can be measured. And, absorbing materials located between the sample and the detector can reduce the number arriving at the detector; and (3) some of the gamma rays will pass through the detector without interacting, especially those at very high energy. In other cases, some gamma rays lose part of their energy through Compton scattering and pair production effects, thus contributing to the gamma continuum beneath the peaks of interest; and (4) detector dead-time causes some events not to be measured, during high counting rates, because for a portion of the time the analyzer is busy processing previous signals.

Peak identification is accomplished by consulting a number of compilations of decay schemes and related tables listing the gamma-ray energies and their associated branching ratios (Erdtmann and Soyka 1979, Firestone et al. 1996). The peak area is determined by summing the total number of counts under the peak and subtracting the background. The measured activity for a particular radioisotope is related to the peak area divided by the counting time and corrected for branching ratio and sample-to-detector efficiency.

Calculating Element Concentrations

For NAA measurements using thermal and epithermal neutrons, an equation that expresses the activity present at any time is given by:

$$A = \left(\frac{m}{M} N_A \theta\right)(\varphi_{th}\sigma_{th} + \varphi_{epi}I)P_\gamma \varepsilon SDC, \tag{8.9}$$

where m = mass of sample (g), M = atomic weight (g mol^{-1}), N_A = Avogadro's number (6.02×10^{23} molecules mol^{-1}), θ = isotopic abundance, P_γ = branching ration of the measured gamma ray, ε = efficiency of the detector at the energy of the measured gamma ray, S = irradiation factor $(1 - e^{-\lambda T_i})$, D = decay factor $(e^{-\lambda T_d})$ and C = count factor $(1 - e^{-\lambda T_c})$.

An equation used to calculate the mass of an element present in an unknown sample relative to a standard with known mass of the same element is

$$\frac{A_{sam}}{A_{std}} = \frac{m_{sam}\left(e^{-\lambda T_d}\right)_{sam}}{m_{std}\left(e^{-\lambda T_d}\right)_{std}}, \tag{8.10}$$

where A = the activity or count rate for the sample (sam) and standard (std), m = mass of the element and T_d = the decay time. When performing short irradiations, the irradiation, decay, and counting times are usually identical for all samples and standards (i.e., the steps are performed on individual samples and standards sequentially and all counting uses the same geometry) such that all time-dependent and geometric factors will cancel. Thus the previous equation simplifies to

$$c_{sam} = c_{std} \frac{W_{std}}{W_{sam}} \frac{A_{sam}}{A_{std}},\qquad (8.11)$$

where c = concentrations of the element of interest in the sample and standard, respectively, and W = the weights of the sample and standard.

Comparisons Between XRF and NAA and Their Application to Obsidian Geochemistry

NAA and XRF are both very powerful methods for elemental analysis of materials. Table 8.1 highlights many of the important differences between these techniques with respect to the analysis of geological and archaeological materials such as obsidian.

Table 8.1 A comparison between X-ray fluorescence and neutron activation analysis

	XRF	NAA
Availability	Many lab-based XRF facilities	Nuclear reactor is required
	Number of portable units is increasing rapidly	Number of locations is very limited
	No special facilities needed	Special training in handling of radioactive materials is required
	Minimal training requirements	Radioactive waste is produced
Sample requirements	Preparation is minimal to none	Encapsulate samples in clean containers
	Nondestructive	Slightly destructive
Analytical	Surface analysis (mostly)	Bulk analysis
	10–15 Elements	25–30 Elements
	Sensitivity at parts per million levels for best elements	Sensitivity at parts per million and parts per billion levels for most elements
	Rapid turnaround	Days or weeks may be necessary to complete analysis
	Good accuracy	Excellent accuracy
	Good precision	Excellent precision
	Due to matrix effects, multiple standards are required for a good calibration	Single standard can be used for calibration
	Sample area >0.75 cm^2	Sample weight >5 mg
Interlaboratory comparison	Good to excellent	Excellent
	Depends on equipment and calibration methods used	Consistent and reliable between NAA labs, if the same standard(s) were used
Analytical cost	Standard rates: \$25–45/sample	Standard rates: \$100/sample
	Subsidized rates: \$25/sample	Subsidized rates: \$25–40/sample

With specific regard to studies of obsidian, both analytical techniques are capable of measuring several of the incompatible elements which have proven useful for differentiating between obsidian sources. The incompatible elements are those that have difficulty entering the cation sites of minerals in volcanic magma and, instead, have higher concentrations in the liquid phase of the magma. Two groups of elements that have difficulty entering the solid phase are the light-ion lithophile elements (LILE) and the high-field strength elements (HFSE). The LILE are elements with large ionic radius, such as K, Rb, Cs, Sr, Ba, REEs, Th, and U. The HFSE includes elements with large ionic valences such as Zr, Nb, Hf, and Ta. The amounts of LILE and HFSE elements present in different obsidian sources are dependent on physical properties such as initial composition of the magma, thermodynamic properties experienced by the magma (i.e., pressures, temperatures, partitioning coefficients), and the age of the magma all of which contribute to each obsidian source having a unique composition.

Standardization Methods Used for XRF and NAA

In order to obtain the most accurate and precise data for geological samples by XRF and NAA, many laboratories prefer to grind their samples into fine powders which are then homogenized. For XRF, the powders are typically pressed into uniformly shaped pellets. The XRF spectral data for each sample are compared to a mass-of-element vs. count-rate calibration curves for each element established by analyzing a series of powdered standard reference materials (SRMs) previously prepared and analyzed as pressed pellets in a manner identical to the unknowns.

NAA can use these same powdered standards, but the unknown samples do not necessarily need to be ground into powders unless the material is inhomogeneous. A mass-of-element vs. count-rate calibration curve is also determined from measurements on the NAA standards, but the curve is linear over a greater range of energies and count rates since there are fewer interferences and matrix effects about which one needs to be concerned. In addition, the interferences that occur when using NAA are well understood and easy to correct.

When obsidian artifacts or source samples are ground into powders and pressed into pellets, the processes of grinding and pelletizing the samples will introduce a small amount of contamination from the tools employed. For instance, a tungsten carbide grinding vessel will introduce contamination from W, Co, and Ta; stainless steel grinding vessels will introduce contamination from Fe, Cr, Mn, and Ni; a ceramic grinding vessel will introduce Si and Al contamination; and an agate grinding vessel will introduce Si contamination. Ideally for archaeological purposes, the minimum amount of sample preparation necessary is preferred such that contamination issues will not confuse interpretation of the data.

   The aforementioned SRMs can be purchased as powders from several international sources, including the United States Geological Survey (USGS), National Institute of Standards and Technology (NIST), Institute for Reference Materials and Measurements (IRMM), South Africa Bureau of Standards (SABS), Canadian Certified Reference Materials Project (CCRMP), Geological Survey of Japan (GSJ), and others. The rock powders have been homogenized and concentrations of elements have been tested and certified analytically by the issuing organization using multiple analytical techniques from a number of different laboratories. Unfortunately, for obsidian characterization research the number of obsidian rock standards available for use in analytical work is very limited and all that are known to exist are present in powdered form.

   Clearly, the availability of a number of solid obsidian calibration standards capable of covering a wide range of element concentrations and free of contamination would be desirable for archaeological laboratories interested in performing nondestructive XRF on obsidian artifacts. To make this possible, a collection of source samples from more than forty different obsidian sources around the World covering a wide range of element were analyzed by NAA at the MURR. The NAA data were examined to identify highly homogeneous sources and to obtain the best possible NAA data for the elements.

   The NAA data for the obsidian sources were calibrated relative to certified concentration data for the well known SRM-278 Obsidian Rock available from NIST (see also Graham et al. 1982). In addition, several other international rock standards were analyzed as unknowns to serve as quality control checks on the analyses (see Glascock and Anderson 1993 and also unpublished data from MURR). These standards included: AVG-1 (Andesite), BCR-1 (Basalt), RGM-1 (Rhyolite Glass Mountain, CA), SRM-688 (Basalt Rock), SRM-1633a (Coal Fly Ash), JA-1 (Japanese Andesite), JB-2 (Japanese Basalt), and JR-1 (Japanese Rhyolite). The quality control data from NAA were very consistent with coefficients of variations (CVs) ranging from 1 to 3% relative to the certified means for the highest-sensitivity elements (i.e., Ce, Co, Cs, Dy, Eu, Fe, Hf, K, La, Lu, Mn, Na, Nd, Rb, Sb, Sc, Sm, Ta, Tb, Th, and Yb) and with CV on the order of 4–10% for the less sensitive elements (i.e., Ba, Cl, Sr, U, Zn and Zr) from NAA.

   The obsidian source samples analyzed by NAA and selected for use as solid standards for the XRF calibration were prepared from a number of golf-ball sized rock samples from known source samples that were analyzed first by NAA to prove source homogeneity. The rock samples were trimmed with a high-speed rock saw until cylindrically-shaped samples with a diameter of about 2.5 cm were produced. The cylinders were sawed flat on one end such that they could be glued with a strong epoxy into a cup-shaped plastic holder with interior diameter 2.5 cm, outer diameter of 3.2 cm, and depth of 1.0 cm. The portion of the rock extending beyond the surface of the holder was removed using a precision saw to produce a flat glassy surface for presentation to almost any XRF. The holders with the obsidian inside appear somewhat like small hockey pucks. Samples with cracks or phenocrysts visible to the naked eye were rejected as calibration sources. Three to six pucks were made for each obsidian source.

One of the most significant advantages of NAA over XRF is that the calibration line for NAA is linear over a much larger span of count rates than XRF and fewer standards are required to produce a reliable calibration. On the other hand, XRF requires many more standards with more variable compositions in order to calibrate and to correct for possible matrix effects (see Chap. 2).

Analytical Methodology

NAA Procedures

Samples for NAA are prepared by extracting a number of fragments of about 25 mg size from the interiors of the source samples after moderate crushing with a Carver Press. The small fragments are inspected under a magnifier to eliminate fragments with possible metallic streaks, crush fractures, or large phenocrysts. In general, about 100 mg of obsidian fragments are combined to make the short irradiation samples and about 250 mg are combined to make the long irradiation samples. The short and long NAA samples are placed inside clean, high-purity vials made of polyethylene and quartz, respectively. Weights are recorded to the nearest 0.01 mg. Standards made from SRM-278 Obsidian Rock powder are weighed into the same irradiation vials.

The samples and standard in polyethylene vials are irradiated for 5 s in a thermal neutron flux of 8×10^{13} n cm^{-2} s^{-1} at the MURR. The short-irradiation samples are allowed to decay for 25 min before starting a 12-min measurement. The measurement permits determination of six short-lived elements: Ba, Cl, Dy, K, Mn, and Na. The long irradiation samples in quartz vials are irradiated for up to 70 h using a thermal neutron flux of 5×10^{13} n cm^{-2} s^{-1}. After the long irradiation, the samples are measured twice. The first measurement uses 30 min on each sample and takes place between 7 and 8 days after the end of irradiation to measure seven medium-lived elements: Ba, La, Lu, Nd, Sm, U, and Yb; and the second measurement for 2.5 h per sample is performed between 4 and 5 weeks after the end of irradiation to measure fifteen long-lived elements: Ce, Co, Cs, Eu, Fe, Hf, Rb, Sb, Sc, Sr, Ta, Tb, Th, Zn, and Zr. Although it is possible to measure barium using either a short- or long-irradiation, the data measured after long-irradiation is usually found to be superior.

XRF Procedures

X-ray fluorescence measurements are performed on solid obsidian samples using an *Elva-X* EDXRF spectrometer. The spectrometer is a table-top ED-XRF operated by a personal computer. The spectrometer is equipped with an air-cooled tungsten target anode X-ray tube with 140 μm Be window and thermoelectrically cooled Si-PIN diode detector. The detector has a resolution of 180 eV at 5.9 keV. The beam dimensions are 3 × 4 mm. The X-ray tube is operated at 40 kV using a current-adjusted, count rate of about 6,000 counts per second (about 25% deadtime). These conditions allow eleven elements to be measured in most samples: K, Ti, Mn, Fe, Zn, Ga, Rb, Sr, Y, Zr, and Nb. Measurement times on the XRF are 180 s. Peak deconvolution and calculations of element concentrations are accomplished in the personal computer using the *Elva-X* spectral analysis package. Mass absorption normalization on the *Elva-X* uses a fixed-energy range portion of the background continuum where there are no peaks.

The instrument was calibrated previously by analyzing the suite of forty well-characterized obsidian source samples in the MURR reference collections. NAA data was used to calibrate the elements in the XRF that are in common to both NAA and XRF (i.e., K, Mn, Fe, Zn, Rb, Sr, and Zr). In order to calibrate the elements not possible by NAA, the forty calibration sources were circulated to several other laboratories for a round-robin analytical exercise. Data reported by Steven Shackley from the Berkeley Archaeological XRF Laboratory had the best agreement with our and we used his data for elements that we could not measure by NAA (i.e., Ti, Ga, Y, and Nb). As a result, we were able to establish very satisfactory mass-of-element-to-count-rate calibration curves for our XRF spectrometer.

The final XRF calibration curve installed in the *Elva-X* was then used to analyze portions of obsidian remaining from the central Mexico source samples. The source samples were previously analyzed by NAA at MURR. In most cases, source samples with a freshly exposed, clean surface were used. Beyond possible cleaning of the sample surfaces with a wet paper towel, no other special preparation was performed on the samples. However, samples with visible cracks or phenocrysts were avoided.

In the next section, the original central Mexico obsidian source characterization study is briefly described. A tabulation of the data collected using NAA and XRF is presented along with a discussion of the data comparison.

Characterizing the Obsidian Sources in Central Mexico

Studies of acquisition, exchange, and utilization of obsidian from sites and sources central Mexico have been the focus of geochemical investigations by XRF and NAA for several decades (Boksenbaum et al. 1987; Cobean et al. 1971, 1991; Glascock et al. 1988; Jack and Heizer 1968; Hester 1972; Pires-Ferriera 1975). Because the earliest researchers were in a rush to identify the sources of artifacts from archaeological sites, they did not conduct intensive surveys of sources. When they did visit the sources, they usually collected a small number of source samples from a single deposit. As a result, they were unable to assess the possibility of internal variations within sources. Furthermore, their analytical methods were limited by the lack of available standards. Thus, the early analytical data was entirely qualitative and only useful to the individual researcher. The certified standard reference material SRM-278 Obsidian Rock did not become available from NIST until after 1980. The first NAA work to use certified standards was published by Boksenbaum et al. (1987). However, this work was flawed by mis-identification of gamma-ray peaks and other errors. As a consequence, almost all of the early XRF and NAA data for the obsidian sources in central Mexico was so unreliable that it was impossible to exchange data between laboratories or between different techniques. A lengthy discussion of the problems associated with the earliest obsidian data from central Mexico is described in Glascock et al. (1998).

In order to remedy the above mentioned deficiencies, a systematic survey and collection of the obsidian sources in central Mexico was conducted in the early 1980s by Robert Cobean and James Vogt. During the fieldwork phase of the project, 818 obsidian source samples (weighing a total of 710 kg) were collected from primary outcrops and secondary deposits throughout central Mexico. Source sam-ples were collected in the states of Guanajuato, Queretaro, Michoacan, Hidalgo, Mexico, Puebla, and Veracruz (Fig. 8.6). The obsidian source samples were sent to MURR for analysis by NAA in order to create a comprehensive obsidian source database. The samples were analyzed relative to the certified standard SRM-278 Obsidian Rock which was widely available by this time. The NAA results were reported earlier by Cobean et al. (1991) and Glascock et al. (1988, 1994, 1998).

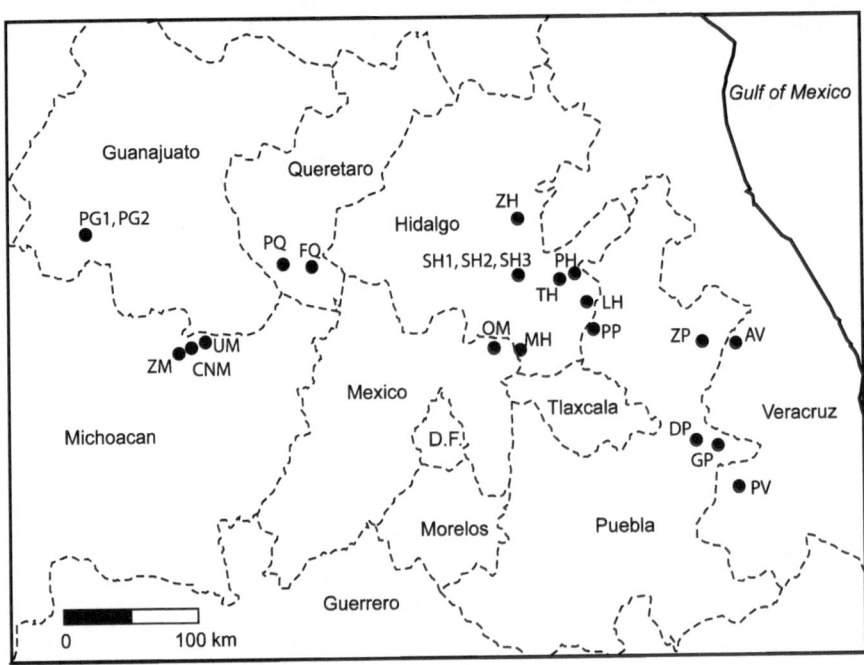

Fig. 8.6 Map of obsidian sources in Mexico

Table 8.2 Names of obsidian sources located in central Mexico and the abbreviations used to identify them on the map in Fig. 8.6	Source name	Map code
	Altotonga, Veracruz	AV
	Zaragoza, Puebla	ZP
	Derrumbadas, Puebla	DP
	Guadalupe Victoria, Puebla	GP
	Pico de Orizaba, Veracruz	PV
	Paredon, Puebla	PP
	Otumba, Estado de Mexico	OM
	Santa Elena, Hidalgo	LH
	Malpais, Hidalgo	MH
	Tepalzingo, Hidalgo	PH
	Tulancingo, Hidalgo	TH
	Zacualtipan, Hidalgo	ZH
	Sierra de Pachuca-1, Hidalgo	SH1
	Sierra de Pachuca-2, Hidalgo	SH2
	Sierra de Pachuca-3, Hidalgo	SH3
	Ucareo, Michoacan	UM
	Zinapecuaro, Michoacan	ZM
	Cerro Negra, Michoacan	CNM
	Fuentezuelas, Queretaro	FQ
	El Paraiso, Queretaro	PQ
	Penjamo-1, Guanajuato	PG1
	Penjamo-2, Guanajuato	PG2

Table 8.2 lists the names for the 22 geochemical source groups from central Mexico and the abbreviations used to show their location on the map in Fig. 8.6. A majority of the central Mexico obsidian sources are located in the states of Hidalgo and Puebla where volcanic activity was quite extensive and obsidian utilization was high. Two of the major sources were found to subdivide into more than one chemical group. The obsidian samples from the Sierra de Pachuca, Hidalgo source were found to subdivide into three geochemical groups; and, the obsidian samples from the Penjamo, Guanajuato source split into two geochemical types. A monograph by Cobean (2002) describes all of the sources in central Mexico in greater detail, and it also discusses their utilization during prehistoric times.

With MURR's acquisition of the *ElvaX* spectrometer in 2006, a decision was made to reanalyze as many of the central Mexico source samples as possible with the new instrument. As shown below, the combined NAA-XRF database makes it possible for researchers to use either analytical technique to calibrate their analytical equipment to obtain data comparable to that from MURR. Today, because the geochemistry of obsidian sources located central Mexico region is known so well, obsidian artifacts (both large and small) from this region are being assigned to sources with nearly 100% success.

Table 8.3 Element concentration means and standard deviations for obsidian sources in central Mexico

Element	Altotonga, Veracruz	Zaragoza, Puebla	Derrumbadas, Puebla	Guadalupe Victoria, Puebla	Pico de Orizaba, Veracruz
NAA data	(n = 14)	(n = 31)	(n = 5)	(n = 23)	(n = 53)
Ba	94 ± 7	451 ± 15	1002 ± 20	931 ± 36	724 ± 42
La	40.1 ± 0.4	37.2 ± 1.2	19.4 ± 1.4	13.7 ± 0.6	6.39 ± 0.95
Lu	0.55 ± 0.02	0.52 ± 0.03	0.055 ± 0.005	0.18 ± 0.02	0.19 ± 0.02
Nd	26.5 ± 0.9	23.4 ± 1.9	13.7 ± 2.1	10.1 ± 1.5	4.89 ± 1.20
Sm	5.55 ± 0.15	5.30 ± 0.11	3.19 ± 0.12	2.37 ± 0.06	1.94 ± 0.07
U	5.7 ± 0.4	5.4 ± 0.4	3.29 ± 0.11	4.6 ± 0.2	5.0 ± 0.2
Yb	3.53 ± 0.06	3.55 ± 0.11	0.40 ± 0.01	1.18 ± 0.07	1.24 ± 0.06
Ce	76.3 ± 1.0	73.1 ± 1.3	39.2 ± 2.6	27.3 ± 1.0	14.2 ± 1.7
Co	0.29 ± 0.01	0.59 ± 0.02	0.15 ± 0.02	0.13 ± 0.01	0.07 ± 0.02
Cs	4.53 ± 0.05	4.04 ± 0.08	4.82 ± 0.07	3.68 ± 0.05	4.01 ± 0.07
Eu	0.185 ± 0.003	0.44 ± 0.01	0.64 ± 0.01	0.35 ± 0.01	0.23 ± 0.01
Fe (%)	0.793 ± 0.011	0.928 ± 0.017	0.856 ± 0.025	0.426 ± 0.006	0.355 ± 0.012
Hf	5.32 ± 0.06	5.76 ± 0.13	2.61 ± 0.08	2.72 ± 0.07	2.40 ± 0.08
Rb	145 ± 2	133 ± 3	110 ± 3	91 ± 2	100 ± 2
Sb	0.60 ± 0.01	0.54 ± 0.04	0.23 ± 0.01	0.23 ± 0.02	0.24 ± 0.01
Sc	2.62 ± 0.03	2.85 ± 0.05	1.20 ± 0.02	1.71 ± 0.02	1.80 ± 0.03
Sr	<10	17 ± 4	136 ± 9	67 ± 6	23 ± 8
Ta	1.68 ± 0.02	1.51 ± 0.03	0.98 ± 0.02	0.80 ± 0.01	0.89 ± 0.02
Tb	0.79 ± 0.01	0.76 ± 0.02	0.30 ± 0.01	0.29 ± 0.01	0.30 ± 0.01
Th	21.1 ± 0.2	19.1 ± 0.4	6.0 ± 0.3	7.64 ± 0.15	6.24 ± 0.24
Zn	38 ± 1	39 ± 1	56 ± 1	27 ± 1	25 ± 3
Zr	128 ± 8	175 ± 8	65 ± 3	54 ± 9	32 ± 7
Cl	1084 ± 217	698 ± 52	547 ± 32	615 ± 113	490 ± 69
Dy	4.77 ± 0.17	4.61 ± 0.23	1.49 ± 0.05	1.84 ± 0.22	1.94 ± 0.20
K (%)	4.03 ± 0.16	4.14 ± 0.15	3.47 ± 0.27	3.39 ± 0.17	3.48 ± 0.21
Mn	238 ± 9	245 ± 8	401 ± 8	518 ± 13	557 ± 14
Na (%)	2.82 ± 0.08	2.91 ± 0.08	3.20 ± 0.06	3.27 ± 0.09	3.18 ± 0.07
XRF data	(n = 12)	(n = 14)	(n = 5)	(n = 15)	(n = 17)
K (%)	3.85 ± 0.13	3.88 ± 0.15	3.41 ± 0.07	3.41 ± 0.06	3.56 ± 0.07
Ti	662 ± 122	933 ± 45	798 ± 84	641 ± 87	543 ± 90
Mn	105 ± 16	135 ± 20	302 ± 51	357 ± 35	333 ± 35
Fe (%)	0.82 ± 0.03	0.92 ± 0.04	0.77 ± 0.02	0.46 ± 0.02	0.42 ± 0.02
Zn	41 ± 5	42 ± 5	43 ± 1	28 ± 3	24 ± 3
Ga	15 ± 1	16 ± 1	13 ± 1	12 ± 1	12 ± 1
Rb	142 ± 4	131 ± 3	113 ± 2	91 ± 3	98 ± 3
Sr	3.4 ± 1.6	32 ± 2	163 ± 4	69 ± 5	31 ± 8
Y	34 ± 3	31 ± 3	2.7 ± 2.2	11 ± 2	12 ± 1
Zr	155 ± 13	186 ± 15	63 ± 6	73 ± 5	58 ± 6
Nb	22 ± 4	21 ± 5	13 ± 2	13 ± 3	14 ± 3

(continued)

Table 8.3 (continued)

Element	Paredon, Puebla	Otumba, Estado de Mexico	Santa Elena, Hidalgo	Malpais, Hidalgo	Tepalzingo, Hidalgo
NAA data	*(n = 28)*	*(n = 37)*	*(n = 10)*	*(n = 19)*	*(n = 10)*
Ba	59 ± 9	761 ± 15	77 ± 17	783 ± 11	892 ± 13
La	53.9 ± 0.7	27.1 ± 0.3	52.1 ± 1.1	25.7 ± 0.4	64.9 ± 0.5
Lu	0.82 ± 0.02	0.33 ± 0.01	0.80 ± 0.03	0.29 ± 0.01	1.04 ± 0.01
Nd	39.6 ± 0.9	18.4 ± 0.5	36.7 ± 2.0	16.7 ± 0.6	67.4 ± 3.3
Sm	7.78 ± 0.13	3.66 ± 0.05	7.53 ± 0.19	3.43 ± 0.06	13.1 ± 0.1
U	4.48 ± 0.30	3.08 ± 0.24	4.52 ± 0.46	3.19 ± 0.14	2.35 ± 0.22
Yb	5.41 ± 0.11	2.22 ± 0.07	5.45 ± 0.20	2.01 ± 0.08	7.22 ± 0.08
Ce	110 ± 2	52.0 ± 0.7	105 ± 2	49.7 ± 0.8	138 ± 2
Co	0.26 ± 0.01	0.65 ± 0.04	0.25 ± 0.01	0.38 ± 0.03	0.16 ± 0.04
Cs	5.45 ± 0.08	3.74 ± 0.07	5.39 ± 0.08	5.10 ± 0.07	4.64 ± 0.04
Eu	0.215 ± 0.003	0.54 ± 0.01	0.22 ± 0.01	0.43 ± 0.01	1.75 ± 0.01
Fe (%)	0.847 ± 0.112	0.865 ± 0.015	0.786 ± 0.011	0.736 ± 0.011	1.83 ± 0.02
Hf	7.36 ± 0.13	4.07 ± 0.05	6.68 ± 0.11	3.32 ± 0.04	12.7 ± 0.2
Rb	159 ± 2	117 ± 2	160 ± 4	115 ± 2	116 ± 2
Sb	1.32 ± 0.02	0.34 ± 0.01	1.27 ± 0.04	0.56 ± 0.02	1.13 ± 0.02
Sc	2.43 ± 0.04	2.11 ± 0.03	2.25 ± 0.03	1.76 ± 0.02	3.70 ± 0.04
Sr	<10	128 ± 9	<10	81 ± 7	60 ± 9
Ta	2.97 ± 0.04	1.13 ± 0.02	2.95 ± 0.04	1.11 ± 0.01	2.07 ± 0.03
Tb	1.18 ± 0.02	0.50 ± 0.01	1.14 ± 0.02	0.47 ± 0.01	1.99 ± 0.04
Th	16.9 ± 0.3	10.3 ± 0.1	16.7 ± 0.3	10.3 ± 0.1	11.1 ± 0.1
Zn	55 ± 1	40 ± 1	50 ± 3	37 ± 1	146 ± 1
Zr	193 ± 8	138 ± 7	178 ± 16	96 ± 4	486 ± 14
Cl	1,191 ± 90	558 ± 45	1,079 ± 253	510 ± 66	983 ± 82
Dy	7.48 ± 0.27	3.08 ± 0.18	7.48 ± 0.56	3.14 ± 0.16	12.0 ± 0.4
K (%)	4.09 ± 0.21	3.41 ± 0.21	4.04 ± 0.11	3.34 ± 0.193	3.47 ± 0.20
Mn	359 ± 9	383 ± 11	354 ± 12	420 ± 11	487 ± 13
Na (%)	2.89 ± 0.07	2.97 ± 0.09	2.88 ± 0.08	3.09 ± 0.09	3.54 ± 0.10
XRF data	*(n = 16)*	*(n = 18)*	*(n = 10)*	*(n = 16)*	*(n = 10)*
K (%)	3.94 ± 0.08	3.54 ± 0.14	3.98 ± 0.12	3.43 ± 0.09	3.62 ± 0.19
Ti	787 ± 94	1,059 ± 45	707 ± 101	783 ± 76	1,587 ± 92
Mn	191 ± 46	335 ± 84	168 ± 44	253 ± 54	334 ± 62
Fe (%)	0.87 ± 0.02	0.84 ± 0.03	0.82 ± 0.02	0.72 ± 0.02	1.69 ± 0.13
Zn	53 ± 4	40 ± 5	50 ± 4	35 ± 5	125 ± 8
Ga	16 ± 1	16 ± 1	16 ± 1	14 ± 1	23 ± 1
Rb	155 ± 2	122 ± 3	159 ± 5	119 ± 2	117 ± 2
Sr	2.3 ± 1.2	155 ± 5	2.2 ± 1.6	99 ± 5	76 ± 2
Y	44 ± 6	20 ± 3	43 ± 3	18 ± 2	64 ± 8
Zr	213 ± 17	143 ± 11	186 ± 11	104 ± 8	448 ± 35
Nb	41 ± 4	15 ± 4	38 ± 3	14 ± 3	41 ± 5

(continued)

Table 8.3 (continued)

Element	Tulancingo, Hidalgo	Zacualtipan, Hidalgo	Pachuca-1, Hidalgo	Pachuca-2, Hidalgo	Pachuca-3, Hidalgo
NAA data	*(n = 40)*	*(n = 20)*	*(n = 129)*	*(n = 11)*	*(n = 27)*
Ba	756 ± 25	252 ± 12	31 ± 12	36 ± 16	29 ± 10
La	74.7 ± 1.5	52.7 ± 0.5	38.6 ± 0.9	63.0 ± 2.2	42.1 ± 2.5
Lu	1.26 ± 0.02	0.70 ± 0.03	1.85 ± 0.04	1.24 ± 0.03	1.62 ± 0.03
Nd	77.9 ± 5.8	33.2 ± 1.6	33.0 ± 2.4	66.3 ± 3.2	43.9 ± 3.4
Sm	15.9 ± 0.3	8.36 ± 0.11	9.90 ± 0.22	15.2 ± 0.4	12.7 ± 0.4
U	3.0 ± 0.7	11.5 ± 0.4	6.8 ± 2.1	4.2 ± 1.0	6.1 ± 2.3
Yb	8.82 ± 0.15	4.91 ± 0.08	12.3 ± 0.3	8.63 ± 0.23	11.2 ± 0.2
Ce	160 ± 3	108 ± 2	92 ± 2	146 ± 5	106 ± 5
Co	0.04 ± 0.01	1.20 ± 0.14	0.05 ± 0.01	0.15 ± 0.01	0.05 ± 0.01
Cs	5.76 ± 0.09	15.7 ± 0.3	3.92 ± 0.06	2.01 ± 0.03	3.08 ± 0.08
Eu	1.65 ± 0.05	0.47 ± 0.02	1.59 ± 0.03	2.50 ± 0.07	1.44 ± 0.03
Fe (%)	1.79 ± 0.03	1.06 ± 0.04	1.58 ± 0.02	1.40 ± 0.03	1.41 ± 0.03
Hf	18.2 ± 0.4	7.08 ± 0.14	27.0 ± 0.4	18.1 ± 0.5	22.0 ± 0.7
Rb	122 ± 3	278 ± 4	192 ± 3	117 ± 2	160 ± 4
Sb	1.76 ± 0.11	1.04 ± 0.03	0.26 ± 0.02	0.14 ± 0.02	0.25 ± 0.02
Sc	0.74 ± 0.02	3.08 ± 0.06	3.21 ± 0.04	4.38 ± 0.24	3.37 ± 0.10
Sr	<10	26 ± 5	<10	<10	<10
Ta	2.32 ± 0.04	1.92 ± 0.02	4.87 ± 0.07	3.10 ± 0.03	4.59 ± 0.11
Tb	2.47 ± 0.06	1.18 ± 0.03	2.25 ± 0.06	2.55 ± 0.09	2.55 ± 0.05
Th	11.8 ± 0.2	35.9 ± 0.5	17.9 ± 0.3	10.8 ± 0.1	16.8 ± 0.4
Zn	175 ± 7	38 ± 5	191 ± 12	120 ± 4	140 ± 9
Zr	736 ± 22	169 ± 10	888 ± 40	694 ± 31	750 ± 64
Cl	1220 ± 92	575 ± 72	1457 ± 152	1082 ± 120	819 ± 124
Dy	16.1 ± 0.9	7.24 ± 0.32	15.8 ± 0.8	15.7 ± 0.8	17.5 ± 1.1
K (%)	3.70 ± 0.23	4.44 ± 0.22	3.78 ± 0.24	4.16 ± 0.18	3.99 ± 0.30
Mn	421 ± 22	170 ± 9	1148 ± 25	790 ± 17	884 ± 49
Na (%)	3.62 ± 0.16	2.45 ± 0.07	3.80 ± 0.09	3.88 ± 0.06	3.60 ± 0.08
XRF data	*(n = 15)*	*(n = 12)*	*(n = 17)*	*(n = 8)*	*(n = 8)*
K (%)	3.71 ± 0.16	4.19 ± 0.15	3.82 ± 0.17	3.95 ± 0.16	3.87 ± 0.10
Ti	1188 ± 71	1099 ± 78	945 ± 144	1379 ± 136	1094 ± 71
Mn	302 ± 28	150 ± 21	797 ± 64	469 ± 65	485 ± 60
Fe (%)	1.73 ± 0.11	1.04 ± 0.05	1.59 ± 0.12	1.36 ± 0.12	1.31 ± 0.06
Zn	169 ± 11	44 ± 4	207 ± 19	139 ± 15	147 ± 7
Ga	22 ± 1	18 ± 2	23 ± 2	20 ± 1	20 ± 1
Rb	122 ± 3	273 ± 4	189 ± 3	115 ± 4	154 ± 8
Sr	15 ± 3	48 ± 4	10 ± 4	<1	<1
Y	87 ± 6	54 ± 3	108 ± 19	82 ± 8	88 ± 10
Zr	687 ± 47	227 ± 22	957 ± 62	707 ± 52	742 ± 40
Nb	50 ± 5	21 ± 5	84 ± 8	57 ± 5	73 ± 5

(continued)

Table 8.3 (continued)

Element	Ucareo, Michoacan	Zinapecuaro, Michoacan	Cerro Negra, Michoacan	Fuentezuelas, Queretaro	El Paraiso, Queretaro
NAA data	*(n = 48)*	*(n = 24)*	*(n = 9)*	*(n = 11)*	*(n = 20)*
Ba	151 ± 26	33 ± 14	25 ± 5	52 ± 14	37 ± 16
La	33.9 ± 1.9	21.9 ± 0.8	10.2 ± 0.7	54.7 ± 1.1	54.9 ± 1.4
Lu	0.36 ± 0.01	0.44 ± 0.03	0.72 ± 0.02	1.31 ± 0.02	2.33 ± 0.07
Nd	23.2 ± 1.7	19.1 ± 1.6	13.3 ± 1.3	57.4 ± 1.7	64.0 ± 2.8
Sm	4.71 ± 0.12	4.91 ± 0.09	5.69 ± 0.13	15.3 ± 0.8	22.0 ± 1.0
U	4.1 ± 0.3	5.6 ± 0.4	7.8 ± 0.4	4.5 ± 0.4	5.1 ± 1.1
Yb	2.36 ± 0.10	3.06 ± 0.14	4.82 ± 0.12	10.3 ± 0.9	17.2 ± 1.2
Ce	65.6 ± 3.5	45.9 ± 1.5	26.7 ± 1.8	125 ± 2	142 ± 4
Co	0.25 ± 0.02	0.12 ± 0.01	0.03 ± 0.01	0.03 ± 0.01	0.02 ± 0.01
Cs	6.88 ± 0.24	9.39 ± 0.33	14.2 ± 0.2	3.02 ± 0.05	3.23 ± 0.07
Eu	0.18 ± 0.01	0.094 ± 0.007	0.027 ± 0.003	0.52 ± 0.01	0.49 ± 0.01
Fe (%)	0.742 ± 0.016	0.685 ± 0.010	0.637 ± 0.006	1.44 ± 0.04	1.95 ± 0.03
Hf	4.12 ± 0.10	4.16 ± 0.07	5.30 ± 0.05	17.2 ± 0.3	32.2 ± 0.7
Rb	144 ± 4	177 ± 5	249 ± 3	170 ± 3	220 ± 7
Sb	0.46 ± 0.04	0.66 ± 0.05	0.68 ± 0.08	0.10 ± 0.03	0.16 ± 0.01
Sc	2.48 ± 0.04	2.78 ± 0.06	3.02 ± 0.03	0.246 ± 0.002	0.155 ± 0.003
Sr	12 ± 3	<10	<10	<10	<10
Ta	1.19 ± 0.04	1.62 ± 0.06	3.24 ± 0.04	2.11 ± 0.03	3.75 ± 0.06
Tb	0.60 ± 0.02	0.73 ± 0.06	1.20 ± 0.03	2.49 ± 0.07	4.39 ± 0.24
Th	13.8 ± 0.3	16.0 ± 0.4	19.8 ± 0.2	20.0 ± 0.3	29.9 ± 0.7
Zn	34 ± 1	38 ± 1	58 ± 3	144 ± 4	234 ± 14
Zr	121 ± 12	75 ± 12	112 ± 42	605 ± 21	1110 ± 48
Cl	429 ± 75	550 ± 45	439 ± 30	836 ± 39	1538 ± 125
Dy	3.75 ± 0.19	4.62 ± 0.33	8.7 ± 1.4	16.4 ± 1.1	29.4 ± 1.1
K (%)	3.92 ± 0.17	3.83 ± 0.20	3.56 ± 0.16	3.74 ± 0.18	3.67 ± 0.21
Mn	167 ± 3	186 ± 6	237 ± 8	230 ± 6	234 ± 7
Na (%)	2.77 ± 0.04	2.85 ± 0.09	3.14 ± 0.10	3.31 ± 0.09	3.61 ± 0.09
XRF data	*(n = 16)*	*(n = 15)*	*(n = 8)*	*(n = 8)*	*(n = 11)*
K (%)	3.81 ± 0.11	3.74 ± 0.09	3.67 ± 0.14	3.69 ± 0.12	3.65 ± 0.20
Ti	575 ± 99	448 ± 19	285 ± 25	673 ± 56	606 ± 81
Mn	96 ± 8	86 ± 4	87 ± 3	211 ± 6	347 ± 15
Fe (%)	0.75 ± 0.01	0.71 ± 0.01	0.67 ± 0.02	1.29 ± 0.04	1.96 ± 0.14
Zn	34 ± 4	34 ± 2	45 ± 1	129 ± 7	244 ± 20
Ga	15 ± 1	14 ± 1	14 ± 1	19 ± 1	28 ± 1
Rb	140 ± 3	170 ± 4	241 ± 6	164 ± 6	218 ± 4
Sr	11 ± 1	2.8 ± 1.9	<1	<1	3.2 ± 0.5
Y	25 ± 2	33 ± 2	57 ± 5	94 ± 5	193 ± 11
Zr	118 ± 12	111 ± 11	131 ± 20	614 ± 20	1247 ± 92
Nb	18 ± 5	21 ± 3	36 ± 3	36 ± 2	69 ± 6

(continued)

Table 8.3 (continued)

Element	Penjamo-1, Guanajuato	Penjamo-2, Guanajuato
NAA data	*(n = 12)*	*(n = 15)*
Ba	80 ± 21	43 ± 26
La	56.0 ± 0.9	45.3 ± 0.8
Lu	1.16 ± 0.01	1.18 ± 0.03
Nd	56.5 ± 2.1	49.4 ± 1.5
Sm	13.5 ± 0.5	12.5 ± 0.4
U	4.7 ± 0.3	5.1 ± 0.5
Yb	8.0 ± 0.1	8.3 ± 0.2
Ce	120 ± 2	101 ± 3
Co	0.02 ± 0.01	0.02 ± 0.01
Cs	5.10 ± 0.09	5.25 ± 0.15
Eu	0.93 ± 0.01	0.80 ± 0.02
Fe (%)	1.62 ± 0.02	1.39 ± 0.04
Hf	16.2 ± 0.3	16.5 ± 0.5
Rb	150 ± 2	151 ± 5
Sb	0.35 ± 0.02	0.29 ± 0.02
Sc	1.31 ± 0.02	0.97 ± 0.03
Sr	<10	<10
Ta	1.74 ± 0.03	1.75 ± 0.04
Tb	1.96 ± 0.09	2.10 ± 0.09
Th	14.4 ± 0.2	14.1 ± 0.4
Zn	130 ± 7	121 ± 3
Zr	588 ± 15	603 ± 16
Cl	1080 ± 77	1096 ± 86
Dy	13.5 ± 0.5	13.8 ± 0.5
K (%)	3.73 ± 0.17	3.77 ± 0.20
Mn	400 ± 12	327 ± 8
Na (%)	3.43 ± 0.07	3.43 ± 0.08
XRF data	*(n = 12)*	*(n = 12)*
K (%)	3.56 ± 0.17	3.68 ± 0.12
Ti	800 ± 110	654 ± 125
Mn	261 ± 51	203 ± 7
Fe (%)	1.57 ± 0.15	1.32 ± 0.06
Zn	128 ± 15	121 ± 8
Ga	20 ± 1	20 ± 1
Rb	143 ± 2	150 ± 3
Sr	4.3 ± 1.1	2.7 ± 1.0
Y	83 ± 6	86 ± 4
Zr	609 ± 44	612 ± 28
Nb	35 ± 7	32 ± 3

Results

A total of 596 of the original 818 obsidian samples collected by Cobean and Vogt from sources of central Mexico were analyzed by NAA with 27 elements determined using the procedures described earlier. XRF was used to analyze 275 of the samples previously studied by NAA. All 22 of the geochemical groups from central Mexico were represented among the samples analyzed. A summary of the analytical results is presented in Table 8.3 where the means and standard deviations from NAA and XRF are listed for each geochemical group. The NAA results appear in the top portion of the table and the XRF results are located at the bottom. Seven elements (i.e., K, Mn, Fe, Zn, Rb, Sr, and Zr) were measured by both techniques in most of the samples. However, the element Sr was not observed by NAA in samples with concentrations below the approximate detection limit of 10 ppm for NAA. And, although the element Mn was measured in every sample by both techniques,

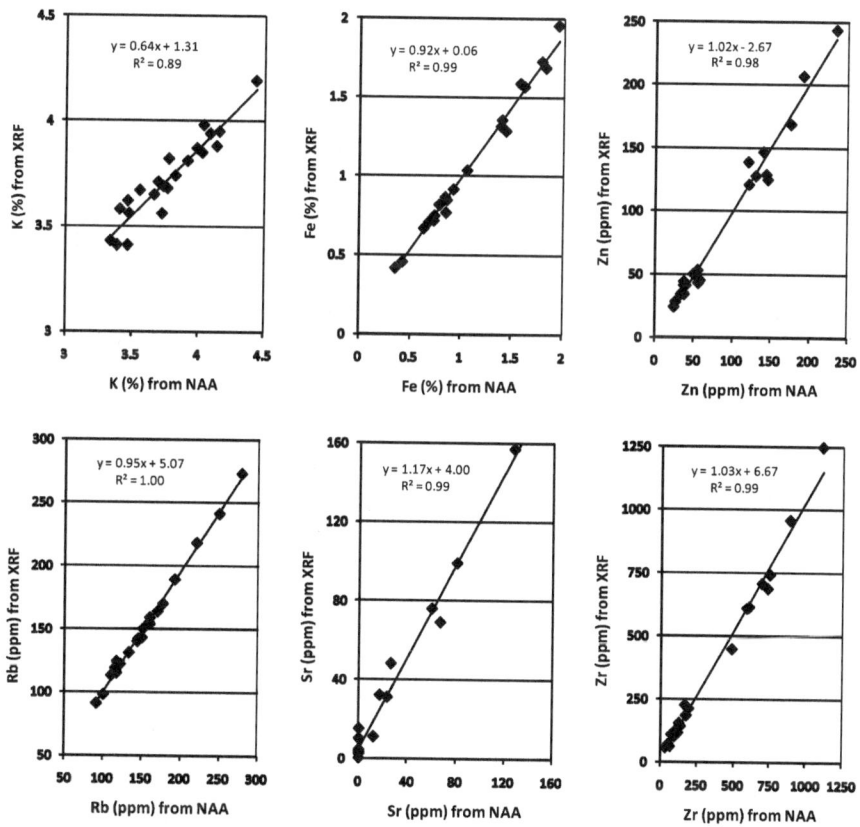

Fig. 8.7 Plots of XRF vs. NAA for selected elements from obsidian means from central Mexico

the XRF data for Mn are much less reliable due to the presence of the small Mn peak on the tail of the large peak from Fe.

An inspection of the data in Table 8.3 suggests that all 22 of the source groups are reasonably homogeneous. More than half of the elements measured have CVs below 3% (NAA elements: Ce, Cs, Eu, Fe, Hf, La, Mn, Na, Rb, Sc, Sm, Ta, Th and Yb; and XRF elements: Fe and Rb). The remaining elements have CVs ranging from 4 to 10%.

In Fig. 8.7, plots show a comparison between the calculated group means by NAA and XRF for K, Fe, Zn, Rb, Sr, and Zr respectively. For five of the six plots, the results show excellent linearity (R^2 is 0.99) when comparing the NAA vs. XRF data for the elements common to both techniques except K. The lower quality data for K are probably a result of the limited concentration range (between 3 and 4%) and the effects of sample shape and surface differences which affect the low-energy X-rays for K more seriously than the other elements. The results are impressive and they support our stated goal of being able to compare with confidence the compositional data for solid samples of obsidian measured using either XRF or NAA.

Discussion

Before beginning to analyze artifacts and assigning the artifacts to specific sources, there are several questions we must answer: (1) Have all of the possible sources been identified? (2) Which information within the database differentiates sources from one another such that artifacts can be assigned a unique provenance? (3) Do we have to perform the XRF, long-NAA, and short-NAA experiments to measure all elements on every sample before assigning provenance? (4) Or, can we measure certain elements and still be able to assign provenance with confidence? The answers to these questions can be found by uncovering patterns within the data by techniques that are often referred to as *data mining*.

Data Mining

In order to use the information in the central Mexico obsidian compositional database for the purpose of assigning artifacts to specific sources, we need to identify elements or groups of elements that enable each source to be uniquely differentiated from every other source. The requirements for sourcing artifacts are best explained by the provenance postulate (Weigand et al. 1977) which requires us to identify one or more elements that have less variation with a source than the variation between sources. Or, in other words the sources must not overlap on all compositional parameters. The effort becomes increasing more difficult when there are many sources and many elements to consider. Some sources may differ on many elements and others may be different on the basis of a single element. Obviously, the greater the number of elements we measure that show a source to be different from other sources, the greater confidence we will have in the accuracy of source assignments made to artifacts when using the data for these elements.

The usual methods to establish source assignments are: (1) to examine the differences between sources one element at a time; (2) to examine a series of bivariate plots; or (3) to use multivariate methods such as cluster analysis, principal components or discriminant analysis. When there are many sources and a large amount of data to be examined, the first method is quite tedious. Bivariate plots are a very useful tool for sourcing obsidian artifacts. If the number of possible sources in the comparison is low, a single bivariate plot may be the most satisfactory solution. When the number of possible sources is high, then examination of several bivariate plots is usually necessary. Multivariate methods are very powerful because they allow one to use all of the information, including correlations between elements, but they are sometimes more complicated than necessary for obsidian. On the other hand, one of the uses for multivariate methods can be to estimate the probability that a misassignment can occur (Glascock et al. 1998). Because bivariate plots are the most expedient and most popular method for making source assignments, they are the method that we will employ here.

Figures 8.8–8.11 show bivariate plots of the obsidian source data using different pairs of elements measured by NAA or XRF. In order to reduce clutter on the plots, the individual samples are not shown. Instead, the compositional groups are displayed in the form of shaded ellipses representing the distribution of geochemical

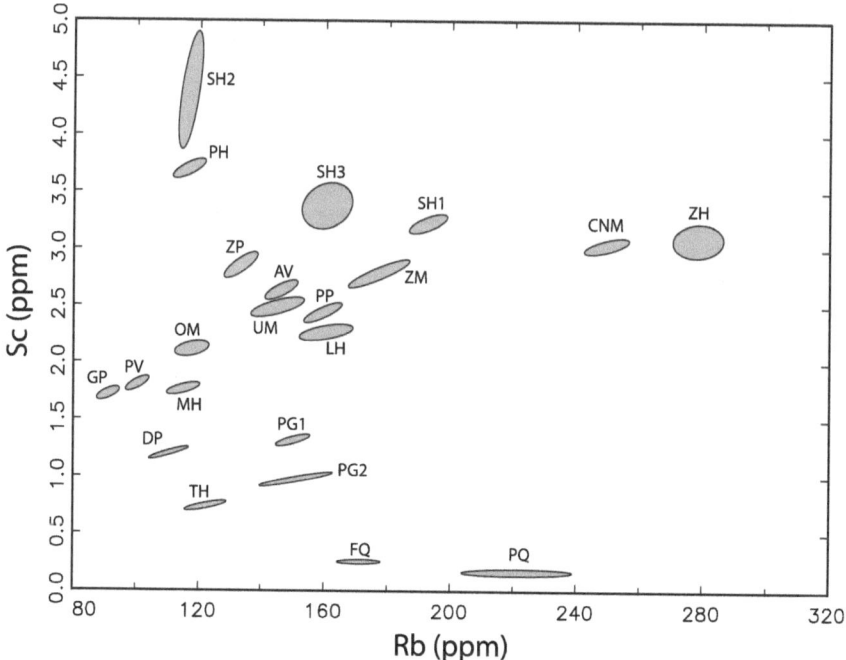

Fig. 8.8 Plot of long-NAA data for Rb and Sc for geochemical source groups located in central Mexico. Individual sources are represented by confidence ellipses calculated at the 95% interval

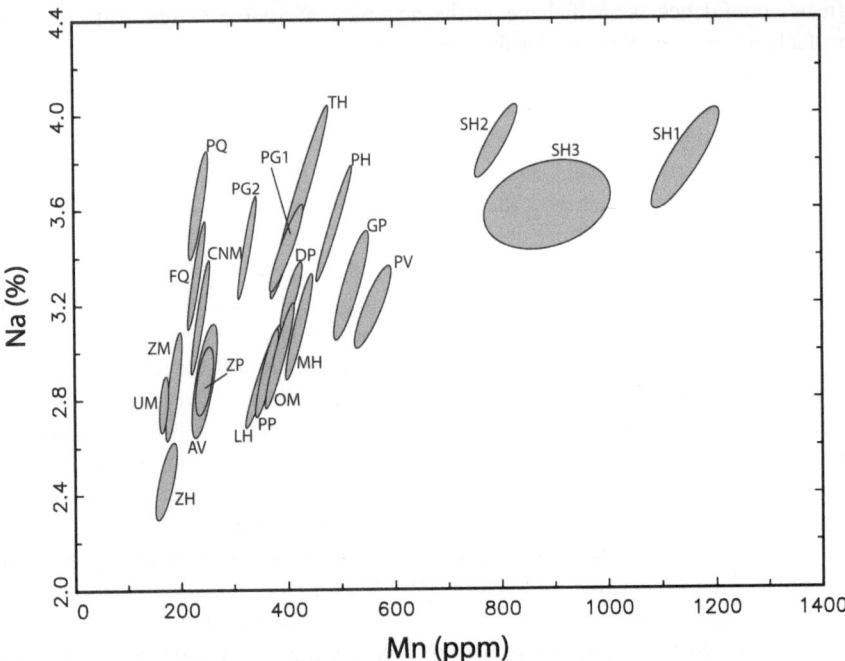

Fig. 8.9 Plot of short-NAA data for Mn and Na for geochemical source groups located in central Mexico. Individual source groups are represented by confidence ellipses calculated at the 95% interval

groups on the pair of elements plotted. The ellipses are plotted at the 95% confidence level which means that 95% of the source samples plot inside the ellipse and fewer than 5% of the samples plot outside the limits of the ellipse. Most of the samples not inside the ellipse will be near the ellipse. The elliptical shapes also help to indicate the degree of correlation between the pair of elements. The three plots have been selected because they are best plots for summarizing the overall differentiation between the sources in central Mexico.

Figure 8.8 displays a bivariate plot of the long-NAA data for the elements Cs and Sc on which all of the geochemical source groups are separated at the 95% level or greater without overlaps between source ellipses. In addition, by examining the data in Table 8.3, we find that the single most critical element is Sc which in this study is the only element showing a significant difference between the Paredon (PP) source with Sc = 2.43 ± 0.04 ppm and Santa Elena (LH) source with Sc = 2.25 ± 0.03 ppm. Scandium is one of the elements only possible on samples after a long irradiation. Thus, when required to distinguish samples between the PP and LH sources, the long-NAA experiment must be performed.

Figure 8.9 presents a bivariate plot of the short-NAA data for elements Mn and Na. This analytical procedure is also referred to as the abbreviated-NAA method in Glascock et al. (1994). Many of the source groups differ by amounts greater than the 95% confidence ellipses, including the subsources at Sierra de Pachuca (SH1,

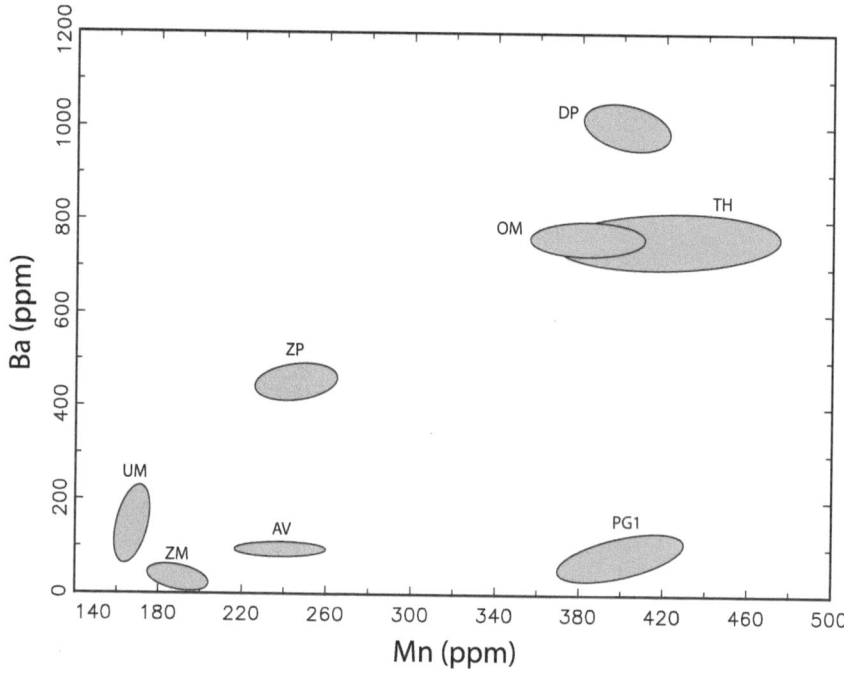

Fig. 8.10 Plot of short-NAA data for Mn and Ba for selected geochemical source groups located in central Mexico. Individual source groups are represented by confidence ellipses calculated at the 95% interval

SH2, and SH3) and Penjamo (PG1 and PG2). However, several pairs of sources overlap at the 95% confidence level, including Ucareo (UM) and Zinapecuaro (ZM), Altotonga (AV) and Zaragoza (ZP), Penjamo-1 (PG1) and Tulancingo (TH), Santa Elena (LH) and Paredon (PP), and Derrumbadas (DP) and Otumba (OM). Fortunately, several of these sources can be differentiated by using the data for two other elements Ba and Dy also measured by the short-NAA procedure. As shown in Fig. 8.10 which plots the elements Ba and Mn, the shaded ellipses representing the sources at UM and ZM, AV and ZP, PG1 and TH, and DP and OM do not overlap when we use the element Ba. Therefore, the only sources not differentiated by the short-NAA procedure are Paredon and Santa Elena which we showed above are only different on the element scandium.

Figure 8.11 shows a bivariate plot of the XRF data of the confidence ellipses for source groups based on Rb and Zr. Although a majority of the source ellipses are separated at the 95% confidence level or greater, there are four clusters of sources showing an overlap. The overlapping sources are: (1) SH2 with TH; (2) PG1, PG2, FQ and SH3; (3) AV with UM; and (4) LH with PP. Some of the overlapping sources can be differentiated using the XRF data for Fe. For instance, the mean concentrations for Fe are 1.36 ± 0.12 % for the SH2 source group and 1.73 ± 0.11

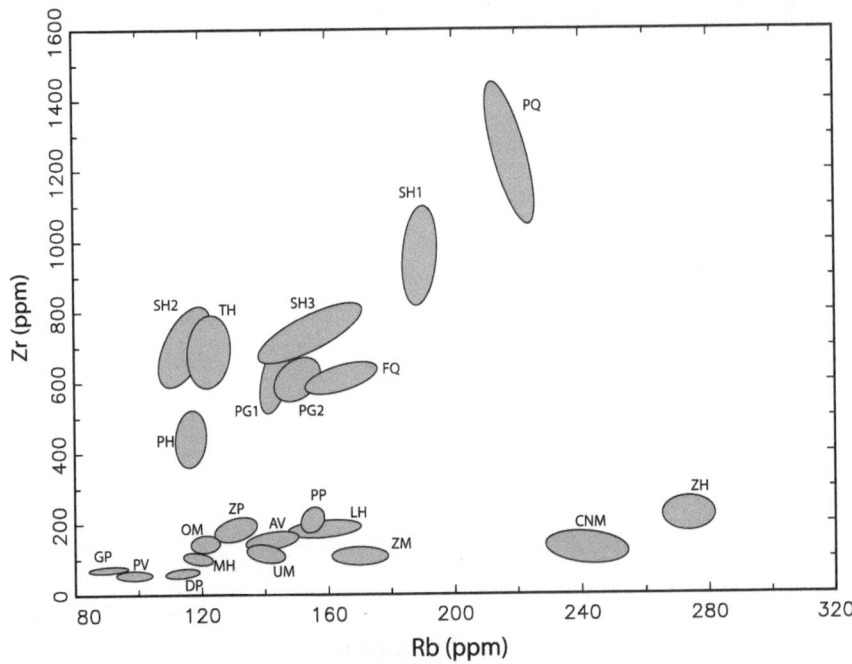

Fig. 8.11 Plot of XRF data for Rb and Zr for geochemical source groups located in central Mexico. Individual source groups are represented by confidence ellipses calculated at the 95% interval

% for the TH source group. The differences and their standard deviations are sufficiently separated for these sources do not overlap at 95% on Fe.

For the PG1 and PG2 sources, mean concentrations for Fe by XRF are 1.57 ± 0.15 % and 1.32 ± 0.06 %, respectively. However, the Fe concentrations for these sources are not sufficiently different from SH3 (Fe = 1.31 ± 0.06 %) and FQ (Fe = 1.29 ± 0.04%) to discriminate the PG2 subsource from the SH3 and FQ sources. The simplest solution is to use the short-NAA data for Mn as shown in Fig. 8.9 and on Table 8.3. In this instance, all three sources FQ (Mn = 230 ± 6 ppm), SH3 (Mn = 884 ± 49 ppm), and PG2 (Mn = 327 ± 8 ppm) are different at the 95% level or greater.

The XRF data for AV and UM does not permit the successful differentiation between these sources. However, we can again use the data Fig. 8.9 and Table 8.3 for Mn which uniquely differentiate between the AV (Mn = 238 ± 9 ppm) and UM (Mn = 167 ± 3 ppm) sources. Data for the element Dy also shows significant differences between the AV (Dy = 4.77 ± 0.17 ppm) and UM (Dy = 3.75 ± 0.19 ppm) sources.

Finally, the XRF results for the LH and PP sources on Fig. 8.11 show a high degree of overlap on the elements Rb and Zr which we observed earlier when using the short-NAA data for Mn, Na, and Ba. This supports the earlier conclusion that the only way to differentiate between the Santa Elena and Paredon sources is to use the long-NAA data for Cs and Sc as shown in Fig. 8.8.

A Protocol for Artifact Analysis at MURR

Now that we understand the capabilities and limitations of our central Mexico obsidian source database using the three basic measurements (i.e., long-NAA, short-NAA and XRF), we can design a protocol that minimizes the analytical costs and reduces the turnaround time required to complete an obsidian artifact provenance project. By performing our measurements on artifacts using the procedures in an order reversed from that used to analyze source samples, a more efficient method is possible.

Experiment 1: XRF

The first experiment we use on artifacts from sites in central Mexico is XRF. Because XRF is nondestructive and rapid, the data are available almost immediately. No sample preparation is required and we do not have to schedule a time to irradiate the samples in the reactor. The measurement times are very short such that it is possible to collect data for up to 100 samples in a single day. The artifact data from the XRF can be compared to different bivariate plots such as Rb and Zr, Rb and Sr, Fe and Zr, etc. As we have shown above in Fig. 8.11, most of the sources in central Mexico do not overlap at the 95% confidence level, except: (1) Sierra de Pachuca-2 with Tulancingo; (2) Penjamo-1, Penjamo-1, Fuentezuelas and Sierra de Pachuca-2; (3) Altotonga with Ucareo; and (4) Santa Elena with Paredon. Data for the element Fe and Sr will separate some of these sources as described above, but others combinations require NAA to give a correct source assignment.

Obsidian artifacts not assignable to sources using the XRF results and artifacts that are too small for XRF, should be prepared for short-NAA or long-NAA. Small artifacts can be important for some archaeological studies. As shown by Eerkens et al. (2002), the distribution of sources for small obsidian artifacts can be greatly different from the distribution of sources for larger artifacts, and for a variety of social and technological reasons may indicate greater distance to source and source assemblage diversity than can be detected by XRF.

Experiment 2: Short-NAA

The short-NAA experiment at MURR requires that samples be placed in polyethylene vials for irradiation. The recommended amount of sample for analysis is 100 mg, but smaller samples down to 10–20 mg are possible. The analytical procedure employed at MURR for obsidian analysis by short-NAA allows up to 70 artifacts and standards to be processed in a single day using two detectors operating in parallel.

As shown in Fig. 8.9, the results from short-NAA for Mn and Na are adequate to differentiate between most of the sources in central Mexico. The main exceptions are the overlapping sources: (1) Ucareo with Zinapecuaro; (2) Altotonga with Zaragoza; (3) Penjamo-1 with Tulancingo; (4) Santa Elena with Paredon; and (5) Derrumbadas with Otumba. If any of these are the possible source(s) for our artifact(s), then the concentration data for Ba or Dy may resolve sourcing questions as illustrated by Fig. 8.10. However, there may still be issues with some of the overlapping sources such as Santa Elena and Paredon or outliers that do not agree with any of the XRF or short-NAA source groups. In this event, the samples should be prepared for long-NAA.

Experiment 3: Long-NAA

The long-NAA experiment at MURR requires that samples be placed in quartz vials for irradiation. The recommended amount of sample for analysis is 200–300 mg, but samples weighing as little as 5 mg are possible also. The procedure employed at MURR for obsidian analysis by long-NAA allows up to 70 artifacts and standards to be processed over period of about 1 month using two detectors operating in parallel on two automatic sample changers. Obviously, the required time and expense are greatest for the long-NAA experiment.

As shown in Fig. 8.8 above for the elements Cs and Sc, the long-NAA experiment is the most successful and comprehensive procedure for analysis of obsidian. Since the long-NAA also measures several other useful elements, the long-NAA results make sourcing of artifacts possible with a high-level of confidence. Therefore, the probability of an erroneous assignment using long-NAA data is very low.

Although not demonstrated here, we have previously shown that when multivariate methods are applied to the full NAA data, the probability of assigning samples to the incorrect source are less than 4 in 100,000 for all sources except the LH and PH pair discussed above (Glascock et al. 1998). In this latter case where the choices are LH and PH, the element Sc is the only significantly different element and the chance for a misassignment increases to a maximum of 5/100.

The time required and the cost of performing the long-NAA can be problems for large projects or in cases where funding is limited. The archaeologist must decide the level of confidence necessary for source assignments of artifacts in competition with the amount of funding available to analyze all samples in a project. Clearly some short-cuts that save expense can be made by using the three-step protocol described here.

Conclusions

The advantages and disadvantages of XRF and NAA as analytical methods for provenance studies of homogeneous archaeological materials such as obsidian artifacts have been described. XRF is a rapid, inexpensive, and non-destructive technique but it does not measure as many elements in obsidian as NAA. NAA measures more elements with greater precision than XRF, but NAA requires that a portion of the sample to be sacrificed and is more time consuming and expensive than XRF.

Using the solid obsidian source samples and higher precision data from NAA to develop a series of calibration curves for elements common to both NAA and XRF, we have succeeded in showing that the data for several elements (i.e., K, Fe, Zn, Rb, Sr, and Zr) are consistent and acceptably comparable between the two techniques. This gives high confidence that data on obsidian samples measured by XRF can be compared directly to NAA collected by NAA.

In this study, a comprehensive NAA and XRF database for the obsidian sources in central Mexico has been presented from the analysis of solid obsidian fragments for NAA and larger solid samples for XRF. Examination of the source data with bivariate plots presents a method by which one can first use XRF to analyze obsidian artifacts from central Mexico. In most cases, XRF should be sufficient to determine the provenance of the artifacts. However, if the artifacts come from a source that overlap on elements measured by XRF, then short-NAA and long-NAA are a possible way of obtaining the data critical to sourcing these difficult artifacts. The analytical protocol at MURR used to source obsidian artifacts from sites in central Mexico is: (1) XRF; (2) short-NAA; and (3) long-NAA. The sourcing results obtained are highly reliable and cost-efficient.

Acknowledgments The author wishes to express his appreciation of support to colleagues Robert Cobean and Jeff Ferguson. He also acknowledges the assistance of undergraduate Christopher Oswald who helped to prepare and analyze many of the obsidian samples by XRF. Any errors or omissions in this work are the responsibility of the author. The Archaeometry Lab at MURR is supported in part by a grant from the National Science Foundation (DBS-0802757).

References

Baum, E. M., Knox, H. D., & Miller, T. R. (2002). *Nuclides and Isotopes: Chart of the Nuclides,* 16th edition. New York, Knolls Atomic Power Laboratory.

Boksenbaum, M. W., Tolstoy, P., Harbottle, G., Kimberlin, J., & Neivens, M. D. (1987). Obsidian industries and cultural evolution in the Basin of Mexico before 500 B.C. *Journal of Field Archaeology, 14,* 66–75.

Cobean, R. H. (2002). *A World of Obsidian: The Mining and Trade of Volcanic Glass in Ancient Mexico.* Mexico, University of Pittsburgh Latin American Archaeology.

Cobean, R. H., Coe, M. D., Perry, E. A., Jr., Turekian, K. K., & Kharkar, D. P. (1971). Obsidian trade at Sal Lorenzo Tenochtitlan, Mexico. *Science, 174,* 141–146.

Cobean, R. H., Vogt, J. R., Glascock, M. D., & Stocker, T. R. (1991). High-precision trace-element characterization of major Mesoamerican obsidian sources and further analyses of artifacts from San Lorenzo Tenochtitlan, Mexico. *Latin American Antiquity, 2*(1), 69–91.

Eerkens, J. W., King, J., & Glascock, M. D. (2002). Artifact size and chemical sourcing: studying the potential biases of selecting large artifacts for analysis. *Society for California Archaeology Newsletter, 36*, 25–29.

Erdtmann, G., & Soyka, W. (1979). *The Gamma Rays of the Radionuclides*. Weinheim, Chemie.

Firestone, R. B., Shirley, V. S., Baglin, C. M., Chu, S. Y. F., & Zipkin, J. (1996). *The Table of the Isotopes*, 8th edition. New York, Wiley.

Glascock, M. D. (2008). Archaeometry. In D. M. Pearsall (Ed.), *Encyclopedia of Archaeology* (pp. 489–494). Oxford: Academic.

Glascock, M. D., & Anderson, M. P. (1993). Geological reference materials for standardization and quality assurance of instrumental neutron activation analysis. *Journal of Radioanalytical and Nuclear Chemistry* 174: 229–242.

Glascock, M. D., Neff, H., Stryker, K. S., & Johnson, T. N. (1994). Sourcing archaeological obsidian by an abbreviated NAA procedure. *Journal of Radioanalytical and Nuclear Chemistry, 180*, 29–35.

Glascock, M. D., Braswell, G. E., & Cobean, R. H. (1998). A systematic approach to obsidian source characterization. In M. S. Shackley (Ed.), *Archaeological Obsidian Studies* (pp. 15–65). New York, Plenum.

Glascock, M.D., Elam, J. M., & Cobean, R. H. (1988). Differentiation of obsidian sources in Mesoamerica. In R. M. Farquhar, R. G. V. Hancock, and L. A. Pavlish (Eds.), *Archaeometry 88, Proceedings of the 26th International Archaeometry Symposium* (pp. 245–251). Toronto, University of Toronto.

Graham, C. C., Glascock, M. D., Carni, J. J., Vogt, J. R., & Spalding, T. G. (1982). Determination of elements in the National Bureau of Standards' geological standard reference materials by neutron activation analysis. *Analytical Chemistry, 54*(9), 1623–1627.

Guerra, M. F. (2008). Archaeometry and museums: Fifty years of curiosity and wonder. *Archaeometry, 50*(6), 951–967.

Hester, T. R. (1972). Trace element analysis of obsidian from the site of Cholula, Mexico. *Contributions of the University of California Archaeological Research Facility, 16*, 105–110.

Jack, R. N., & Heizer, R. F. (1968). 'Fingerprinting' of some Mesoamerican obsidian artifacts. *Contributions of the University of California Archaeological Research Facility, 5*, 81–100.

Jenkins, R. (1999). *X-ray Fluorescence Spectrometry*. 2nd edition. New York, Wiley-Interscience.

Pires-Ferriera, J. W. (1975). *Formative Mesoamerican Exchange Networds with Special Reference to the Valley of Oaxaca*. Memoirs of the Museum of Anthropology No. 7. Ann Arbor, University of Michigan.

Sansonetti, J. E., Martin, W. C., & Young, S. L. (2005). *Handbook of Basic Atomic Spectroscopic Data* (version 1.1.2). Available online: http://physics.nist.gov/Handbook. Gaithersburg, National Institute of Standards and Technology.

Weigand, P. C., Harbottle, G., & Sayre, E. V. (1977). Turquoise sources and source analysis: Mesoamerican and the southwestern USA. In T. K. Earle and J. E. Ericson (Eds.), *Exchange Systems in Prehistory* (pp. 15–32). New York, Academic.

Chapter 9
Is There a Future for XRF in Twenty-First Century Archaeology?

Rosemary A. Joyce

I am pretty widely characterized (by others) as a "post-processual" archaeologist, more specifically (an identification I actually agree with), as a feminist archaeologist. So what am I doing, enthusiastically endorsing the idea that the future of archaeology requires us to integrate archaeological science even more fully into our practice and explanations than we have been doing in recent decades in archaeology? Without obscuring my actual lack of direct experience in the application of XRF, which I still admit to treating like a kind of magic, I want to make two arguments in this chapter, explaining why archaeologists like me should encourage the cultivation of expertise in archaeological science, and why archaeological scientists should find what I personally prefer to call "social archaeology," a congenial place to spend time.

Why Archaeological Science Should Be Part of Social Archaeology

There are a number of distinctions we could draw in contemporary archaeology, many of which are primarily products of disciplinary histories that obscure more than they reveal. Conducting graduate level seminars in the history of theory in archaeology is a good way to come to understand that fiercely debated positions that animated the field when current faculty were graduate students seem quaint to present archaeological students, who are developing their ideas in a pluralist theoretical environment. Truth to tell, it has always been hard for me to police the boundaries of the supposed schools of thought in anthropological archaeology because, as a practicing Mesoamericanist, my graduate training politely ignored the

R.A. Joyce (✉)
Department of Anthropology, University of California, Berkeley, CA 94720-3710, USA
e-mail: rajoyce@berkeley.edu

M.S. Shackley (ed.), *X-Ray Fluorescence Spectrometry (XRF) in Geoarchaeology*,
DOI 10.1007/978-1-4419-6886-9_9, © Springer Science+Business Media, LLC 2011

whole critique of culture history that animated Binfordian New Archaeology and went on encouraging collection of types and varieties in the pursuit of assembling sufficient data so that we would be able to write a complete culture history some day. I had to learn that I was a "post-processual" archaeologist from an anonymous reviewer of my dissertation book manuscript, and needed to turn to my then-colleague Bob Preucel to find out that Ian Hodder had, while I was living in Honduras, stopped being known for his work on spatial analysis with Clive Orton and become the leader of a heretical sect accused of denying the reality of the past and saying all explanations were equally good.

So I am inclined to be wary about absolutism in archaeology, and I am happy to be considered eclectic by others. But that does not mean that I see all archaeological work as being based on compatible assumptions. I did once argue that we simply recognize a productive pluralism in contemporary archaeology that started after processualism's period of theoretical dominance was shattered by interpretive, feminist, Marxist, and other archaeologies, uneasily lumped together under the postprocessual label (Preucel 1995). I still think history shows that pluralism is an engine of self-correction and progress within a discipline that we should prize and encourage. But I now can put a name to what I personally think is the strongest current of contemporary archaeological theory, the one I think is emerging as the likely dominant mode of archaeological analysis for the coming generation. And I want to argue that this new way of doing social archaeology needs absolutely the best contemporary archaeological science and is already incorporating it, and influencing the contemporary development of archaeological materials science in positive ways.

As a member of the founding group of editors of a new archaeological journal that began publishing 8 years ago, I signed on a statement arguing for a newly marked "social archaeology" (Journal of Social Archaeology Editors 2001). We defined what we meant by social archaeology in part by citing issues related to which archaeology was making significant contributions, issues that were then of broad interdisciplinary interest: identity, the body, social memory, temporality, diasporas, life experience, ritual, household archaeology, and, most important here, material culture studies. We argued that "an explicit focus on 'the social' in terms of identity, meaning and practice" characterized archaeology, which we defined as "a process of knowledge production mediated by material culture and experience" (Journal of Social Archaeology Editors 2001, p. 6). Our commitment then, and now, was to theoretical pluralism as long as the authors of prospective contributions took "the social" as central in understanding the past. We argued that "the environment, the economy and technologies are fundamentally social" (Journal of Social Archaeology Editors 2001, p. 9). While advocating a theoretically pluralist position, we singled out materiality as one of the areas where archaeology has a particular ability to contribute to interdisciplinary inquiry into "the social" writing that archaeology's "maturity and confidence derives from advances in our understanding of material culture and the long term" and calling for examination of "how material culture is continuously implicated in webs of signification in the processes of creating meaning" (Journal of Social Archaeology Editors 2001, p. 9).

The actual pattern of publication in the *Journal of Social Archaeology* (JSA) since then bears out our argument. As of this writing, over 130 papers published in JSA have material culture or technology as a central topic, and most of them draw on archaeological science to make their arguments. Many of these contributions focus on the nature of materials, and the technologies developed to work them. Thus Demattè (2006) centers her study of the Chinese Neolithic on an extended discussion of the properties of jade and related stones, their procurement, and processing. Other authors use findings from archaeological science to trace "object biographies." A study of the creation of commemorative monuments in Florida draws on the correlation of paddle markings and compositional analysis of ceramic paste to argue for the collection of pottery made in dispersed localities in specific mortuary features (Wallis 2008, pp. 243–244, 255–258).

While many of these authors are not practitioners of the archaeological sciences on which they rely, other writers, in JSA and elsewhere, are creating new data using archaeological materials science as well as engaging in the more theoretical work on materiality that increasingly draws on such materials science. Jones (2004, 2005), e.g., has contributed both to the JSA and to *Archaeometry*. In both venues, he demonstrates the productivity of combining contemporary archaeological science with approaches rooted in what I would call social archaeology (for which his proxy has been "archaeological theorists"). Jones (2004), p. 327 argues that "both groups are fundamentally engaged in the same task: an understanding of past societies through the medium of material culture," a position with which I would concur.

Like many others, Jones (2004), p. 328 identifies a reliance on linguistic models, and a concept of signs as arbitrary, as barriers that blocked the development of productive connections between new ways of doing archaeology in the 1980s and archaeological materials science. With the emergence of archaeologies of meaning grounded in Peircean semiotics in the last decade (e.g., Bauer 2002; Preucel 2005; Preucel and Bauer 2001; Joyce 2007, 2008; Knappett 2002; Lele 2006), social archaeology now counts on theories of meaning that take seriously the motivated nature of signs. Signification, in contemporary social archaeology, is as likely to be based on essential relations between signs and what they represent, including the physical properties of things that signify, or in technical terms, index, the places of origin of things, the social practices through which things were created, or the experiential qualities of things (their durability, softness, color, and the like). Social archaeologists are developing approaches that privilege materiality, rather than treating things as texts, making physical properties of things critical to understanding social life, a position that is also gaining ground in ethnography (e.g., Keane 2008).

From this perspective, the recent history that Jones (2004, pp. 329–330) describes as dividing idealist and empiricist branches of archaeology, his archaeological theorists and archaeological scientists, is ripe for reunion. The subject–object divide that attributes all agency to people (the subjects) has been definitively questioned. Human action and the materials through which humans act cannot be divided into an immaterial essence and a stable externalized set of things. Jones (2004, p. 335) concludes that theoretical archaeologists were late to realize that "it is the very physicality of the archaeological record, in the form of artifacts and architecture, that is the strength of the discipline" calling for analyses that bridge the fields of archaeological materials science and what I would call more precisely social archaeology by examining "not only the description and characterization of the material properties of artifacts (the traditional preserve of archaeometry), but also on how those material properties intervene in the social lives of people (the traditional concern of [social] archaeology)."

We can take this examination of the overlap between recent research in archaeological materials science and social archaeology a bit further, and in the process, wind our way back to the main topic of this book, which is, after all, not all archaeological science but the potential of X-ray fluorescence spectrometry, a specific method most often applied to geoarchaeological materials like obsidian, basalt, and culturally produced materials like ceramics. Where the *Journal of Social Archaeology* has published more than 100 articles drawing substantively on archaeological materials science or dealing with understanding archaeological materials and technologies, a search of *Archaeometry* for the phrase "social archaeology" produced only 11 records for articles published from 1985 to the present, the earliest such articles using compositional analyses to suggest patterns of trade.

Starting in 2000, however, a series of articles began to consider the implications of "technological choice" approaches for archaeological materials science (Sillar and Tite 2000). Suggesting that technologies may be "determined as much by local perceptions and the social context as any material constraints or purely functional criteria," Sillar and Tite (2000, p. 2) illustrate one of the main reasons for convergence between archaeological materials science and social archaeology: both see the project of understanding archaeological materials as based on recognizing the minimal degree of any kind of determinism that might simplify our explanatory challenge. Yes, people needed to cover a basic level of subsistence needs and arrange for biological reproduction to take place, or the societies they formed could not persist over time; but the plasticity of human biology, behavior, and, above all, technology allowed for an extremely wide range of approaches to solving these basic problems of existence, while fostering an incredible degree of activity that went beyond what might be considered strictly necessary under minimizing/maximizing logics. As Sillar and Tite (2002, p. 15) write, "'technological style' or those performance characteristics associated with the cultural framework involve culture-material interactions that are not subject to...universal laws and, thus, tend to be culturally specific."

Technological style or technical choice becomes the core concept of half of the subsequent articles associated with the phrase "social archaeology" published in *Archaeometry*. These topics are staples of social archaeology and its exploration of human technical agency. What this suggests is that archaeological scientists and social archaeologists have begun to discover common ground where human beings through their more or less knowledgeable actions were the agents of production of technologies. From both perspectives, technology, underdetermined by adaptive efficiency, is a subject for empirical examination. What constitutes the empirical today, however, is different than it might have been in the mid-twentieth century: human practice, even intentionality, can no longer simply be set aside.

While still rare, there are other signs on the landscape of convergence between the program I call social archaeology and applications of materials science in archaeology. Pollard and Bray (2007), reviewing the prospects of integration of archaeological science within broader archaeology, suggest that the questions archaeologists are asking today have created new opportunities to engage with archaeological science, specifically citing practice theory, theories of agency, and considerations of the agentive role of things. They write that "this is not a move toward a material determinism; instead it is more a move to a balance where the inherent physical and chemical properties of materials and their geographical distribution also have an active impact on human culture" (Pollard and Bray 2007, p. 253). They specifically mention provenance and life histories of objects, two more of the thematic foci of social archaeological work on materiality.

David Killick (2004), in his contribution to a debate about how to study technology, identifies many of the same topics of social archaeology as common concerns with archaeological materials scientists. Technological style, technological choice, practice theory, agency, materiality, and material culture study, all are among the topics he singled out for discussion. Archaeologists interested in these topics, he argued, would agree that "there is usually more than one technology that satisfies the minimum requirements for any given task" and that "the choice of a particular technology from a pool of satisfactory alternatives may be strongly influenced by the beliefs, social structure, and prior choices" of the people under study, with little interest in "the search for grand unifying theories, or 'master narratives' of technological change" (Killick 2004, p. 571, 572). Because technology is embedded in social relations, Killick argues, "technology in pre-industrial societies creates persons as well as products" (p. 572). Everything here is familiar to social archaeologists, even though in this case, the author is, like Jones, an anthropological materials scientist.

So, I think we have demonstrated that contemporary archaeological materials science benefits from the concepts and programs of contemporary social archaeology. But I want to make a stronger argument, which is that social archaeology needs archaeological materials science. Remember that while 11 articles in *Archaeometry* explicitly concern social archaeology, more than 100 articles in the *Journal of Social Archaeology* concern materiality and may build directly on archaeological materials science. Why would that be? And, finally, what might this imply for the future of XRF in twenty-first century archaeology?

Why Social Archaeology Needs Archaeological Science

It should be obvious that I think archaeological materials science is increasingly drawing on social concepts to organize findings, conceptualize research problems, and create explanations. But social archaeology is making inroads today, I would argue, because social archaeologists are using methods including those of archaeological materials science to produce increasingly fine-grained information from materials. Social archaeologists prize the details that materials science can provide us because these small telling details, and the variability that being attentive to them reveals, are information for us, not just noise. If our goal in archaeology is to find generalizable patterns, then archaeological materials science is often actually unhelpful; the more work that is done on characterization within an assemblage, the more variability we see. But if our goal is to understand human actions in the past that we think were in large part historically contingent, socially varied, and even expedient, then all that variability is interpretable information.

It is probably no surprise that one of the most pervasive uses of archaeological science in social archaeology has been the exploration of bone chemistry as a means to understand the relative status of categories of people. Over time, the more the individual analyses using any method were completed, the more it became clear that the simple assumptions we used to formulate our models needed to be complicated. So, e.g., Christine White (2005) assembled data from multiple studies of a diversity of Maya archaeological sites, all designed to explore gendered differences in diet. Where in each individual study, it might have been possible to make general interpretations, by the time all the samples were assembled, the overarching conclusion was that "gendered dietary differences vary by resource, time, and site location, which is consistent with [the] view that female status and power were unevenly negotiated over time and place" (White 2005, p. 375). The social archaeological research that White was citing my own argument on the fluidity of Maya gender systems, based on entirely different evidence... images of men and women – but it is the archaeological science that shows that what I could see in stereotyped representations was also evident in everyday life.

Social archaeologists, who have used materials analyses to begin to create osteobiographies, have adopted archaeological materials science to explore the particularities of place and everyday practice as well. Cynthia Robin (2002) employed soil chemistry in a modest village to reveal patterns of land use, circulation, and likely activities in which different families would have come together intermittently, creating village-wide social links. As with the example of bone chemistry contributing to understanding gendered lives in all their diversity, this kind of analysis has become central to social archaeology's attempt to understand the complexity of everyday life understood as an engine of social continuity and change. Household archaeology, the subfield of social archaeology transformed by analyses like this, grew out of the examination of regularities of settlement patterns that treated buildings as objects in the way critiqued by Jones (2004) and Killick (2004). Soil chemistry gives social archaeologists the ability to see households in action.

So Where Does XRF Fit in?

Clearly, archaeological science gives social archaeology the kind of data that our theories need: particular to specific objects, providing more ways to see diversity, and providing insight into the dynamics of things. Yet there is something clearly missing in the cases I have cited so far: where, in all this positive view of the future of archaeological science in the twenty-first century, is XRF analysis?

Lithics, principal materials for XRF analyses and the subject of this volume, are still much less common foci of social archaeological analyses. Among research on bone chemistry, soil chemistry, ceramic composition, and iron technology that have provided some of the best examples, there are remarkably few studies of stone tools. Residues identified on chipped stone are one notable exception, providing the basis for a pathbreaking analysis that takes residues as evidence for past actions (Haslam 2006). But the obsidian to which the residues adhered is given relatively little attention, referred to only in terms of presumed raw material source.

Perhaps the most widely cited contribution to social archaeology based fundamentally on the study of lithic materials, analyzed using XRF to determined sources, is Steven Silliman's (2001) study of colonial indigenous workers at the Petaluma adobe, who persisted in using traditional obsidian sources in the nineteenth century. Silliman argues convincingly that the continuity of use of traditional sources was an active approach to maintaining ongoing cultural identity, showing us that humble chipped stone could be powerful evidence for social archaeological models.

So lithic materials analyzed using XRF are being used as evidence today by archaeologists interested in addressing a variety of social questions. But there is more that needs to happen for social archaeologists to conduct research on lithics as central as research on ceramics or other preferred and long-studied materials. My own view is that this late realization of the potential of lithics for social archaeology comes in part from the apparent simplicity of the materials on the surface, which seem to promise less information than the showier technologies like metallurgy and ceramics with their extensive productive installations and stylistic traits. Archaeological materials science has revealed unexpected dimensions to lithics, and social archaeological models of technical choice and object biographies provide an excellent basis on which to integrate these ubiquitous materials into new interpretations.

In a move that directly addresses one of the concerns expressed by both Killick and Jones, Steve Shackley undertakes a clear description of the underlying science (Shackley, Chapter 1). One of the points repeatedly made about archaeological materials science is the need for the work to be done by archaeologists, archaeologists who understand the science well enough to employ the data in appropriate ways. As I said in the beginning of this essay, I still tend to treat XRF as if it is magic; Steve wants to make sure that I, and others like me, cannot get away with this. We have to undertake to understand the science so that we are active participants in the construction of knowledge using analytic results. The science behind XRF is easy to understand and good for us to think about. The inherent variability of materials is something we still are inclined to under-emphasize.

A second point that these articles make is the importance of non-destructive methods in contemporary archaeology. This is a major change from the situation of mid-twentieth century archaeology, probably due in equal parts to the increasing responsibility of archaeology to stakeholders, including descendant groups, for whom cultural heritage concerns make destruction of things unattractive; the strengthening of the conservation ethic in archaeology that has always been based on the understanding that we need to preserve our research materials as much as possible; and the realization with each newly developed method that an opportunity has been lost to apply the method to materials excavated (and thus destroyed) in good faith. XRF should be widely employed if for no other reason than simply because we can analyze samples and still retain them with most of the information they offer intact. (I have to admit that as I write this, obsidian artifacts selected for XRF analysis of composition are waiting in my lab for research to be completed on starches and possible phytoliths adhering to their surfaces. There are some days I wish I could pretend I don't know what potential there is in attempting multiple analyses on these humble artifacts, particularly in the field as we separately bag every obsidian fragment, a far cry from the normative practice when I was trained as a Mesoamerican archaeologist. On the other hand, how extraordinary it is to be able to take one piece of obsidian worked around 1000 BC and trace its journey from a local outcrop of obsidian to a small oval house where it was used to chop root crops.)

Among the many reasons why XRF is a method that social archaeologists should appreciate and adopt, and possibly why it is driving an increasing number of research projects is the development and rapid decrease in price of portable XRF devices. One of the lingering questions that many of us have had about this great leap forward is whether the portable devices actually can produce data of comparable quality to the laboratory-based XRF analyses. Again, in this volume, we have some solid data to guide us in considering this question. Portable XRF devices are opening up the potential to capture data from curated collections that cannot be removed to a lab, unlocking extraordinary potential for social archaeologists to incorporate many more objects in our analyses and thus, potentially, see greater diversity in the material media of past social relations.

New data on new devices are part of rethinking our criteria for selecting methods; a tradition that includes classic papers, which drew our attention to the problem of false precision, once used to argue that Instrumental Neutron Activation Analysis was inherently better as a method than XRF. A major contribution in this volume is the comparison of INAA and XRF in the study of obsidian and basalt. The conclusion, which is that sometimes one method is appropriate, and sometimes another, should be entirely intelligible to any social archaeologist. Of course, it does mean that we need to understand all these methods, and their variants, well enough to choose the appropriate, not merely available or traditional, approach.

So where does XRF fit in? Using this low cost, non-destructive method on an assemblage of obsidian from the Late Classic and Terminal Classic periods at the site of Cerro Palenque, archaeologist Julia Hendon has been able to argue that rather than pursuing a site-wide strategy of minimizing cost, or creating a single monopolistic relationship with a single site providing all the obsidian used at the site, the people of Cerro Palenque obtained obsidian in small quantities from a wide range of sources, many of them quite distant. She argues that for the people of Cerro Palenque, this cosmopolitanism of social relations, mediated through exchange of obsidian, was what was of greater social value. We could build on these observations and on joint work Hendon and I have done together with Jeanne Lopiparo on the experience of place in the Ulúa Valley, where Cerro Palenque grew to be the single largest town in the Terminal Classic period, between 800 and 1000 AD. Based on a variety of analyses, we found that people in the Ulua Valley oriented their ritual actions toward geographic points on the horizon, mountains that were also sources of valued stone resources. In a situation where stones used for items of daily use indexically signified their places of origin, the gathering together of obsidian from many sources transformed Cerro Palenque into a place containing many places. We are at a point where we can look at the differences in distribution within the site of obsidian from these many different sources, and begin to suggest possible gendered patterns of the experience of expanded senses of place, by tracing where obsidian was in use in kitchen contexts, and where it was restricted to ritual deposits while locally available cherts were instead employed. And because of the low cost and non-destructive nature of XRF, we have been able to amass a large body of analyzed samples from sites in the valley dating as early as 1650 BC and through the eighteenth century, when obsidian continued to be used in colonial settings. We are in a position to put together the kind of meta-analysis of patterns that Christine White carried out with bone chemistry, only with many more data points allowing us to track the subtleties of changing choices by makers and users of chipped stone through a historical tradition lasting for millennia. This is the kind of richness that XRF promises archaeology of the twenty-first century, and is already delivering. It is a way of interrogating materiality that can open windows into past social experiences that are limited only by our ability to imagine how materials that XRF can be applied to mediate the lives of people across space and over time. It is an exciting moment in social archaeology and archaeological materials science, specialties that are growing so close that it probably is time to stop worrying about their differences and start building on their shared engagements.

References

Bauer, A. (2002), Is What You See What You Get? Recognizing Meaning in Archaeology. *Journal of Social Archaeology* 2, 37–52.

Demattè, P. (2006), The Chinese Jade Age: Between Antiquarianism and Archaeology. *Journal of Social Archaeology* 6, 202–226.

Haslam, M. (2006), An Archaeology of the Instant? Action and Narrative in Microscopic Archaeological Residue Analysis. *Journal of Social Archaeology* 6, 402–424.

Jones, A. (2004), Archaeometry and Materiality: Materials-Based Analysis in Theory and Practice. *Archaeometry* 46, 327–338.

Jones, A. (2005), Lives in Fragments? Personhood and the European Neolithic. *Journal of Social Archaeology* 5, 193–224.

Joyce, R. A. (2007), Figurines, meaning, and meaning-making in early Mesoamerica. In C. Renfrew and I. Morley, Eds., *Material Beginnings: A Global Prehistory of Figurative Representation*, (pp. 107–116). Cambridge: McDonald Institute for Archaeological Research.

Joyce, R. A. (2008), Practice in and as deposition. In B. Mills and W. Walker, Eds,, *Memory Work*, (pp. 25–40). Santa Fe: School of American Research Press.

Journal of Social Archaeology Editors (2001), Editorial Statement. *Journal of Social Archaeology* 1, 5–12.

Keane, W. (2008), On the Materiality of Religion. *Material Religion* 4, 230–231.

Killick, D. (2004), Social Constructionist Approaches to the Study of Technology. *World Archaeology* 36, 571–578.

Knappett, C. (2002), Photographs, Skeumorphs and Marionettes: some Thoughts on Mind, Agency and Object. *Journal of Material Culture* 7, 97–117.

Lele, V. P. (2006), Material Habits, Identity, Semeiotic. *Journal of Social Archaeology* 6, 48–70.

Pollard, A. M., and Bray, P. (2007), A Bicycle Made for Two? The Integration of Scientific Techniques in Archaeological Interpretation. *Annual Review of Anthropology* 36, 245–259.

Preucel, R. W. (1995), The Postprocessual Condition. *Journal of Archaeological Research* 3, 147–175.

Preucel, R. W. (2005), *Archaeological Semiotics*. Malden: Blackwell.

Preucel, R. W. and Bauer, A. (2001), Archaeological Pragmatics. *Norwegian Archaeological Review* 34, 85–96.

Robin, C. (2002), Outside of Houses: The Practices of Everyday Life at Chan Nòohol, Belize. *Journal of Social Archaeology* 2, 245–268.

Sillar, B., and Tite, M. S. (2000), The Challenge of "Technological Choices" for Materials Science Approaches in Archaeology. *Archaeometry* 42, 2–20.

Silliman, S. (2001), Agency, Practical Politics, and the Archaeology of Culture Contact. *Journal of Social Archaeology* 1, 190–209.

Wallis, N. J. (2008), Networks of History and Memory: Creating a Nexus of Social Identities in Woodland Period Mounds on the Lower St. Johns River, Florida. *Journal of Social Archaeology* 8, 236–271.

White, C. D. (2005). Gendered Food Behaviour Among the Maya: Time, Place, Status and Ritual. *Journal of Social Archaeology* 5, 356–382.

ERRATUM TO:

Elemental Analysis of Fine-Grained Basalt Sources from the Samoan Island of Tutuila: Applications of Energy Dispersive X-Ray Fluorescence(EDXRF) and Instrumental Neutron Activation Analysis (INAA) Toward an Intra-Island Provenance Study

Phillip R. Johnson
Department of Anthropology, Texas A&M University,
College Station, TX 77843-4352, USA
e-mail: phillipjohnson@tamu.edu

M.S. Shackley (ed.), *X-Ray Fluorescence Spectrometry (XRF) in Geoarchaeology*,
DOI 10.1007/978-1-4419-6886-9, pp. 143–160, © Springer Science+Business Media, LLC 2011

DOI 10.1007/978-1-4419-6886-9_10

In chapter 7 on page 155 Figure 7.9 is wrong. The correct figure is:

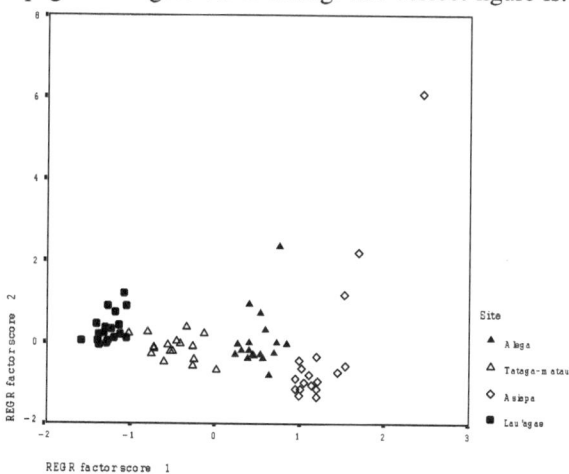

Fig. 7.9 Biplot of first two principal component analysis (PCA) scores from EDXRF data

In chapter 7 on page 155 Figure 7.10 is wrong. The correct figure is:

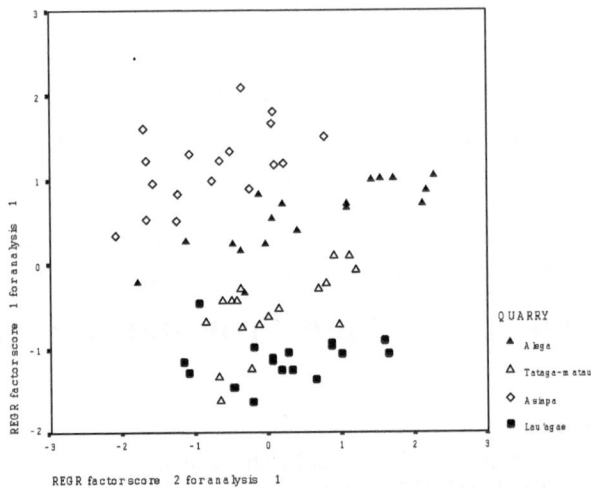

Fig. 7.10 Biplot of first two PCA scores from INAA data

In chapter 7 on page 156 Figure 7.11 is wrong. The correct figure is:

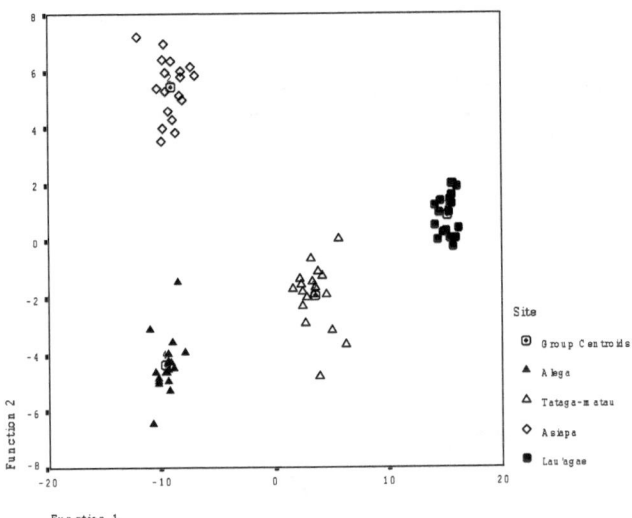

Fig. 7.11 Biplot of first two canonical discriminant analysis (CDA) functions from EDXRF data

In chapter 7 on page 156 Figure 7.12 is wrong. The correct figure is:

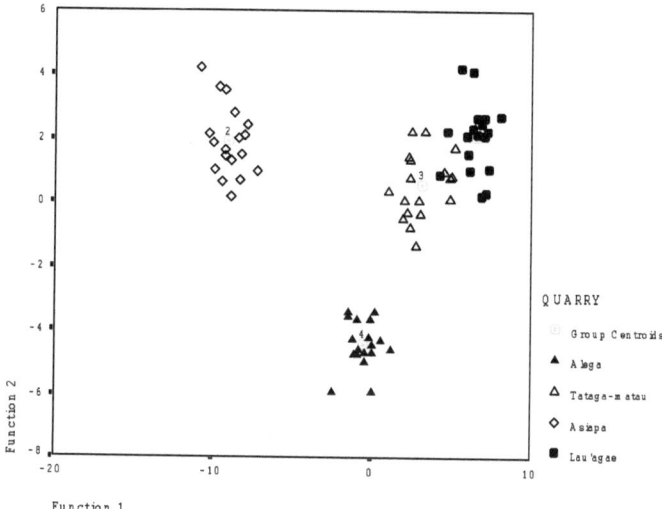

Fig. 7.12 Biplot of first two CDA functions from INAA data

The online version of the original chapter can be found at
http://dx.doi.org/10.1007/978-1-4419-6886-9_7

Appendix

Laboratory Sampling, Analysis and Instrumentation with the Berkeley Thermoscientific Quant'x EDXRF Spectrometer

The following is the analytical trajectory followed at the Berkeley XRF lab. It is substantially similar the other lab protocols for all the reasons discussed in the volume (see also Lundblad et al. 2008).

All archaeological samples are analyzed whole. The results presented are quantitative in that they are derived from "filtered" intensity values rationed to the appropriate X-ray continuum regions through a least squares fitting formula rather than plotting the proportions of the net intensities in a ternary system (McCarthy and Schamber 1981; Schamber 1977). Or more essentially, these data through the analysis of international rock standards, allow for inter-instrument comparison with a predictable degree of certainty (Hampel 1984).

All analyses for this study were conducted on a ThermoScientific *Quant'X* EDXRF spectrometer, located in the Geoarchaeological XRF Laboratory, Department of Anthropology, University of California, Berkeley. It is equipped with a thermoelectrically Peltier cooled solid-state Si(Li) X-ray detector, with a 50 kV, 50 W, ultra-high-flux end window bremsstrahlung, Rh target X-ray tube and a 76 μm (3 mil) beryllium (Be) window (air cooled), that runs on a power supply operating 4–50 kV/0.02–1.0 mA at 0.02 increments. The spectrometer is equipped with a 200 l/min Edwards vacuum pump, allowing for the analysis of lower-atomic-weight elements between sodium (Na) and titanium (Ti). Data acquisition is accomplished with a pulse processor and an analog-to-digital converter. Elemental composition is identified with digital filter background removal, least squares empirical peak deconvolution, gross peak intensities and net peak intensities above background.

For the analysis of mid Zb condition elements Ti-Nb, Pb, Th, the X-ray tube is operated at 30 kV, using a 0.05 mm (medium) Pd primary beam filter in an air path at 200 s livetime to generate X-ray intensity Kα-line data for elements titanium (Ti), manganese (Mn), iron (as $Fe_2O_3^T$), cobalt (Co), nickel (Ni), copper, (Cu), zinc, (Zn), gallium (Ga), rubidium (Rb), strontium (Sr), yttrium (Y), zirconium (Zr), niobium (Nb), lead (Pb), and thorium (Th). Not all these elements are reported since their values in many volcanic rocks are very low. Trace element intensities were converted to concentration estimates by employing a least-squares calibration line rationed to the Compton scatter established for each element from the analysis of international rock standards certified by the National Institute of Standards and Technology (NIST), the US. Geological Survey (USGS), Canadian Centre for Mineral and Energy Technology, and the Centre de Recherches Pétrographiques et Géochimiques in France (Govindaraju 1994). Line fitting is linear (XML) for all elements but Fe where a derivative fitting is used to improve the fit for iron and thus for all the other elements. When barium (Ba) is acquired in the High Zb condition, the Rh tube is operated at 50 kV and 1.0 mA, rationed to the bremsstrahlung region (see Davis et al. 1998). Further details concerning the petrological choice of these elements in Southwest obsidians is available in Shackley (1988, 1995, 2005; also Mahood and Stimac 1990; and Hughes and Smith 1993). Specific standards used for the best fit regression calibration for elements Ti-Nb, Pb, Th, and Ba, include G-2 (basalt), AGV-2 (andesite), GSP-1 (granodiorite), SY-2 (syenite), BHVO-2 (hawaiite), STM-1 (syenite), QLO-1 (quartz latite), RGM-1 (obsidian), W-2 - (diabase), BIR-1 (basalt), SDC-1 (mica schist), TLM-1 (tonalite), SCO-1 (shale), all US Geological Survey standards, BR-1 (basalt) from the Centre de Recherches Pétrographiques et Géochimiques in France, and JR-1 and JR-2 (obsidian) from the Geological Survey of Japan (Govindaraju 1994).

The data from the WinTrace software are translated directly into Excel for Windows software for manipulation and on into SPSS for Windows for statistical analyses when necessary. In order to evaluate these quantitative determinations, machine data were compared to measurements of known standards during each run. RGM-1 or an appropriate standard for the type of rock is analyzed during each sample run for obsidian artifacts to check machine calibration.

References

Davis, M.K., Jackson, T.L., Shackley, M.S., Teague, T., and Hampel, J., (1998), Factors affecting the energy-dispersive x-ray fluorescence (EDXRF) analysis of archaeological obsidian. In M.S. Shackley (Ed.), *Archaeological obsidian studies: method and theory. Advances in archaeological and museum studies 3*, (pp. 159–180). New York: Springer/Plenum.

Govindaraju, K., (1994), 1994 Compilation of working values and sample description for 383 geostandards. *Geostandards Newsletter* 18(special issue), 1–158.

Hampel, J.H., (1984), Technical considerations in x-ray fluorescence analysis of obsidian. In R.E. Hughes (Ed.), *Obsidian studies in the Great Basin*, (pp. 21–25). Berkeley: Contribution of the University of California Archaeological Research Facility 45.

Hughes, R.E. and Smith, R.L., (1993), Archaeology, geology, and geochemistry in obsidian provenance studies. In J.K. Stein J.K. and A.R. Linse (Ed.), *Scale on archaeological and geoscientific perspectives*, (pp. 79–91). Boulder: Geological Society of America Special Paper 283.

Lundblad, S. P., Mills, P. R., and Hon, K., (2008), Analysing archaeological basalt using non-destructive energy-dispersive X-ray fluorescence (EDXRF): effects of post-depositional chemical weathering and sample size on analytical precision. *Archaeometry*, 50, 1–11.

Mahood, G. A. and Stimac, J.A., (1990), Trace-element partitioning in pantellerites and trachytes. *Geochemica et Cosmochimica Acta* 54, 2257–2276.

McCarthy, J.J. and Schamber, F.H., (1981), Least-squares fit with digital filter: a status report. In K.F.J. Heinrich, D.E. Newbury, R.L. Myklebust, and E. Fiori, (Eds.), *Energy dispersive x-ray spectrometry*, (pp. 273–296). Washington, D.C.: National Bureau of Standards Special Publication 604.

Schamber, F.H., (1977), A modification of the linear least-squares fitting method which provides continuum suppression. In T.G. Dzubay, (Ed.), *X-ray fluorescence analysis of environmental samples*, (pp. 241–257). Ann Arbor: Ann Arbor Science.

Shackley, M. S., (1988), Sources of archaeological obsidian in the Southwest: an archaeological, petrological, and geochemical study. *American Antiquity* 53, 752–772.

Shackley, M.S., (1995), Sources of archaeological obsidian in the greater American Southwest: an update and quantitative analysis. *American Antiquity* 60, 531–551.

Shackley, M.S., (2005), *Obsidian: geology and archaeology in the North American Southwest*. Tucson: University of Arizona Press.

Glossary

The references in this volume are a good source for exploring the technical aspects of X-ray fluorescence spectrometry and archaeological stone. The glossary here is a small sample of what are the more critical terms useful in the field. There are a number of concepts in X-ray fluorescence spectrometry that are too complex for simple glossary definitions, and I refer the reader to the index and/or appropriate chapters. Six very useful references, from which the germ of most of these definitions are derived, are recommended for further reading, and Bates and Jackson (1984) or a similar geological dictionary should be on the shelves of all archaeologists (see also Goffer 1996; Jenkins 1999; Sigurdsson 2000; Thorpe and Brown 1985). For obsidian, the dominant material discussed in this volume, I refer you to the glossary in Shackley (2005). See also www.learnxrf.com.

Absolute temperature The fundamental scale used for measuring temperatures in the physical sciences. Expressed in degrees Kelvin which are calculated by adding 273 to measurements in degrees centigrade.

Absorption edge or absorption edge energy The highest or upper limit of the K and K or L and L line energies. For Rb ($Z = 37$) it is 15.1999998 keV (see Fig. 2.2). This is crucial to understand when calculating the rationed region of interest, and peak overlaps in EDXRF.

Absorption spectroscopy A technique of chemical analysis based on the measurement of electromagnetic radiation absorbed by substances. The wavelength or frequency of the radiation absorbed reveals information about the type of radiation absorbing substance of its constituent elements.

Absorption spectrum The spectrum of radiant energy absorbed by any substance. The wavelengths of the absorbed radiation are identical to those of the radiation released.

Abundance The mean concentration of an element in a geochemical reservoir; i.e., the abundance of rubidium in rhyolite glass. The order of abundance of elements in the earth's crust is O, Si, Al.

Accelerator mass spectrometric analysis A physical method based on the combination of a particle accelerator and a mass spectrometer. Useful in determining the nature and number of atoms in a given isotope, such as used in accelerator mass spectrometric radiocarbon (^{14}C) dating (AMS radiocarbon dating).

Accuracy   The degree that experimental measurements are free from errors, or the degree of error in those measurements – how closely a measurement value obtained conforms to the actual value of the sample. In XRF this is generally computed by the linear error in calibration (see also precision).

Acidic   A term applied to those igneous rocks, such as rhyolite and dacite, that contain more than 60% SiO_2, as contrasted with intermediate (andesite) or basic (basalt) rocks. Silicic is a more common modern term for these rocks.

Activation   The process of making a material radioactive, usually by bombarding the substance with nuclear particles such as neutrons in a reactor, as in activation analysis.

Activity   (1) The property of substances to react with other substances; (2) the number of atoms of a radioactive element decaying per unit time.

Adsorption   The attraction of one substance to the surface of another.

Alkalic igneous rocks   Those igneous rocks that contain more sodium than potassium than is average for that rock group.

Alkali feldspar   Sodium or potassium rich feldspar such as sanidine, common in rhyolite.

Alpha particle   A nuclear particle which is emitted during the decay of certain natural isotopes such as thorium; consisting of two protons and two neutrons with a double positive electrical charge.

Alumina   A refractory material composed of aluminum oxide (Al_2O_3). Aluminum oxides are relatively high in rhyolites, typically over 12 wt.%, and are one of the principal glass formers, along with SiO_2.

Analytical chemistry   The study, theory, and techniques (instrumentation) of determining the composition of matter, either qualitatively or quantitatively. Often known as *analysis*.

Andesite   A gray to black volcanic rock with between about 52 and 63 weight percent silica (SiO_2). Andesites contain crystals composed primarily of plagioclase feldspar and one or more of the minerals pyroxene (clinopyroxene and orthopyroxene) and lesser amounts of hornblende. At the lower end of the silica range, andesite lava may also contain olivine. Andesite magma commonly erupts from stratovolcanoes as thick lava flows, some reaching several kilometer in length, such as the San Francisco Peaks in northern Arizona. Andesite magma can also generate strong explosive eruptions to form pyroclastic flows and surges and enormous eruption columns. Andesites erupt at temperatures between 900 and 1,100°C. The extrusive equivalent of diorite. Andesite will grade into dacite exhibiting more alkali feldspar and quartz.

Angle of incidence   The angle made by a beam of radiation incident to a surface with a line perpendicular to that surface, as in the angle of incidence in wavelength X-ray fluorescence spectrometry.

Ångström unit   A unit of measure that is equal to one ten thousandth of a micron (0.00000001 cm; 10^{-8}). Often Anglicized to Angstrom.

Anion   A negative ion – an atom or molecule that has gained one or more electrons and bears a negative charge.

Aphanitic An igneous rock whose particles have a mean diameter of less than 1/16 mm; fine grained.

Aphyric Fine-grained or glassy volcanic rocks with no observable minerals in hand sample.

Archaeometry The study of the applications of the physical and natural sciences to archaeological problems, to include but not limited to: geophysical survey, materials analysis, dating methods, provenance studies – also archaeological science.

Atom The smallest particle of an element that can take part in chemical reactions.

Atomic absorption spectroscopy A spectroscopic technique based on the measurement of radiation absorbed by the atoms of the constituent elements of a material. Characteristic wavelength of the radiation is absorbed and the intensity of the absorption forms the basis for determining the relative proportion of each element in the substance.

Atomic number The number of protons in the nucleus of an element – equal in value to the number of electrons in an atom of the element (see "Z").

Background Incidental signals obtained when measuring physical phenomena which may be confused with those required or desired for actual measurements. In EDXRF, any number of elements may emit characteristic radiation at various wavelengths that "interfere" with the element of interest. This background radiation is "stripped" from the element of interest radiation through a number of generally linear algorithms (see also bremsstrahlung radiation).

Banded Said of a vein, sediment, or other deposit having alternating layers that differ in color or texture and that may or may not differ in mineral composition, e.g., banded Vulture obsidian from central Arizona.

Basalt A hard, often black volcanic rock with less than about 52 weight percent silica (SiO_2). Because of basalt's relatively low silica content, it has a low viscosity (resistance to flow). Therefore, basaltic lava can flow quickly and easily move >20 km from a vent. The low viscosity typically allows volcanic gases to escape without generating enormous eruption columns. Basaltic lava fountains and fissure eruptions, however, still form explosive fountains hundreds of meters tall. Common minerals in basalt include olivine, pyroxene, and plagioclase. Basalt is erupted at temperatures between 1,100 and 1,250°C. The most common volcanic rock on earth.

Basic Igneous rocks that have relatively low silica content, $\approx$45–50%, such as gabbro and basalt. Basic rocks are relatively rich in iron, magnesium, and/or calcium and include most mafic rocks. Cf. silicic.

Batholith A large, generally discordant plutonic mass that has more than 100 km^2 of surface exposure and no known floor; e.g., the Peninsular Range Batholith of northern Baja California and southern California.

Beam The flow of radiation in only one direction.

Binary system A system consisting of two components, such as the $MgO–SiO_2$ system.

Bombardment   The process of directing highly energy particles against a target element. It may either bounce off, or become absorbed by the nucleus and form an entirely new particle.

Boulder   A detached rock mass larger than a cobble with a diameter greater than 256 mm and generally rounded and indicating evidence of transport.

Bragg's law   Derived by the English physicists Sir W.H. Bragg and his son Sir W.L. Bragg in 1913, stated that for every angle of incident radiation, the only wavelength reflected to the detector in wavelength X-ray fluorescence spectrometry is the one that conforms to Bragg's formula: $n\lambda = 2d \sin \theta$, where n is a whole number $1-n$, is the wavelength of the X-ray radiation used; d is a constant characteristic of every crystalline substance (i.e., the X-ray crystal); and is the angle on incidence of the x-radiation on the sample. So, by changing the angle of the crystal, you can select for specific elements of interest. In the Philips PW 2400 at Berkeley, this is all done automatically and any combination of elements can be analyzed. The system changes crystals for the various elements, calculates the overlap of elements within the spectrum and yields results in any form desired: qualitative, ratio, quantitative, graphic.

Bremsstrahlung radiation   Also called continuous or white radiation is produced as the impinging high energy electrons are decelerated by the atomic electrons of the elements making up the specimen and originating in the anode. In EDXRF, the bremsstrahlung scatter is used for ratioing in the higher energies, particularly for acquiring Ba in obsidian and other volcanic analyses (and Chap. 3).

Calc-alkalic series   Those igneous rocks in which the weight percent silica is between 56 and 61 when the weight percent of CaO and K_2O + Na_2O are equal; those igneous rocks containing plagioclase feldspar.

Calcic series   Igneous rocks in which weight percent Si is greater than 61 when the weight percent CaO and K_2O + Na_2O are equal.

Caldera   A large basin-shaped volcanic depression the diameter of which is many times greater than that of the forming vent or vents (e.g., Mule Creek and Valles Calderas in New Mexico).

Calibration   Data, obtained by measuring reference standards or employing fundamental parameters, used by the XRF instrument to create mathematical models for determining the composition of sample materials. Empirical calibration is based on the analysis of standards with known elemental compositions (see Chap. 2). Fundamental standardless calibration is based on mathematical algorithms that describe the physics of the detector's response to pure elements. In this case, the typical composition of the sample must be known, while the calibration model may be verified and optimized by one single standard sample. Standardless fundamental calibrations are much less tedious than empirical calibrations, but often don't yield the accuracy required for geoarchaeological problems.

Cathode   The negative terminal of an electrical source.

Cation   An ion with a positive electric charge.

Centroid The weighted center of the peak, calculated as the energy in eV at which the side of the peak to the left has the same number of counts as the side of the peak to the right of this point.

Chalcedony A mineral which consists of a porous mixture of microscopic crystals of quartz and amorphous hydrated silica in a fibrous structure. Often produced in intermediate and silicic eruptions by the removal of silica by water, as opposed to chert often precipitated from other sediments and having a granular crystalline structure.

Charge A quantity of electricity due to an excess of electrons (negative charge) or a deficiency (positive charge).

Chemical analysis The resolution of materials into their chemical components. The analysis can be classical (gravimetric of volumetric) or more recently instrumental (e.g., X-ray fluorescence spectrometry).

Chert A hard, dense cryptocrystalline secondary sedimentary rock consisting chiefly of interlocking (granular fabric) quartz crystals less than 30 m in diameter. Chert can contain amorphous silica (opal) or chalcedony. Chert occurs principally as nodules or concretions often in marine environments where it is often precipitated through limestone, or less commonly as layered deposits. It occurs in nearly any color or combinations (agate), and in hand sample can look identical to chalcedony.

Closed basin A region draining to a depression from which water escapes only through evaporation.

Coarse grained A sedimentary rock in which the individual constituents are easily seen with the unaided eye.

Cobble A rock fragment between 64 and 256 mm in diameter, between a pebble and boulder in size.

Collimator A small aperture or optical focusing element used to shape and direct X-rays generated by the X-ray source.

Compton scatter Crucial in the EDXRF and WXRF analysis of whole, nondestructive sample analysis, Compton scattering (C) occurs when the incident X-ray photon is deflected from its original path by an interaction with an electron. The electron gains energy and is ejected from its orbital position. The X-ray photon loses energy due to the interaction but continues to travel through the material along an altered path. Since the scattered X-ray photon has less energy, it, therefore, has a longer wavelength than the incident photon, and can be seen near the Rh peak in instruments with Rh targets. The event is also known as incoherent scattering because the photon energy change resulting from an interaction is not always orderly and consistent. The energy shift depends on the angle of scattering and not on the nature of the scattering medium. This rather simple linear relationship is at the core of the ability to analyze nondestructively by XRF (see calibration). In nondestructive EDXRF and WXRF, the elemental peak heights are rationed to the Compton scatter in order to eliminate matrix and size effects (see Chap. 2). First observed by Arthur Compton in 1923 and this discovery led to his award of the 1927 Nobel Prize in Physics.

Consanguinity   The genetic relationship between igneous rocks that are derived from common magmatic origin (e.g., the obsidian chemical groups within the Mule Creek Volcanic Field; Antelope Creek, Mule Mountains, and Mule Creek/ North Sawmill Creek).

Continental crust   Crustal rocks that underlie the continents that range in thickness from about 35 km to as much as 60 km under mountain ranges.

Convection   In a magma chamber in which the central liquid rises while the marginal liquid descends due to the variability in heat.

Continuous radiation   See bremsstrahlung.

Cryptocrystalline   A rock fabric consisting of crystals too small to be discerned under a light microscope.

Count rate   The number of fluoresced X-rays per unit time counted from the sample under measurement.

Dacite   Dacite lava is most often light gray, but can be dark gray to black. Dacite lava consists of about 63–68% silica (SiO_2). Common minerals include plagioclase feldspar, pyroxene, and amphibole. Dacite generally erupts at temperatures between 800 and 1,000°C. It is one of the most common rock types associated with enormous Plinian-style eruptions. When relatively gas-poor dacite erupts onto a volcano's surface, it typically forms thick rounded lava flow in the shape of a dome. It can form glass, but due to lower proportions of silica and aluminum oxides is generally vitrophyric (e.g., O'Leary Peak glass in the San Francisco Volcanic Field).

Deadtime   Amount of time required by the XRF instrument to detect a fluoresced X-ray and process the signal into a pulse. During this interval, other X-ray events cannot be detected or processed. In nondestructive EDXRF large samples have higher deadtime, while smaller samples have lower deadtimes. Optimally, 50% is best for modern detection in EDXRF.

Detector   XRF component that produces output charges (pulses) that are proportional to energy of X-ray photons entering the detector.

Devitrification   Conversion of a glass to crystalline material, as obsidian to perlite through hydration.

Diffraction   Diffraction occurs as waves interact with a regular structure whose repeat distance is about the same as the wavelength. The phenomenon is common in the natural world, and occurs across a broad range of scales. For example, light can be diffracted by a grating having scribed lines spaced on the order of a few thousand angstroms, about the wavelength of light. It happens that X-rays have wavelengths on the order of a few angstroms, the same as typical interatomic distances in crystalline solids. That means X-rays can be diffracted from minerals which, by definition, are crystalline and have regularly repeating atomic structures. When certain geometric requirements are met, X-rays scattered from a crystalline solid can constructively interfere, producing a diffracted beam. In 1912, W. L. Bragg recognized a predictable relationship among several factors (see Bragg's Law).

EDX One of the acronym's for energy dispersive X-ray fluorescence (XRF) spectrometry. See X-ray fluorescence analysis.

Electron An elementary particle which has a negative electric charge. One of the basic constituents of atoms.

Element A substance that cannot be decomposed into another substance except through radioactive decay (e.g., Rb, Sr, Zr, Nb, Ba).

Elemental analysis The determination of the elemental composition of a substance that can be either qualitative (determining the presence of) or quantitative (determining the relative amounts).

Empirical calibration See calibration.

Eruption The ejection of volcanic materials (lava, pyroclasts) onto the earth's surface either through a vent or fissure.

Escape peak A false peak in the spectrum produced by the occasional loss of some photon energy absorbed by the detector due to fluorescence induced in the detector medium.

eV Electron volt; 1 kiloelectron volt (keV) $= 1.60217646 \times 10^{-16}$ J. In EDXRF, the position of the peaks is measured in keV (Rb K1 $= 13.375$ keV (see peaks).

Excitation The process of a displacement of an electron from its normal or ground state.

Explosive Index The proportion of pyroclasts among the total products of a volcanic eruption.

Extrusive Igneous rocks that have been erupted onto the earth's surface, including lava and ash flows.

Fabric The arrangement of the crystal constituents of a rock.

Felsite A generic term applied to any light-colored aphanitic igneous rock, often hypabyssal and intermediate or silicic in composition. In archaeology often applied to any light colored volcanic rock.

Filter A mechanical device (generally a foil) or mathematical technique used to distinguish X-rays fluoresced by materials with similar characteristic energy levels.

Fine-grained An igneous rock whose particles have a mean diameter of less than 1/16 mm. Synonym – *aphanitic*.

Fluorescence The process by which incident electromagnetic radiation induces atomic ionization. As a result of this ionization, electrons from higher energy orbitals drop (cascade) to lower energy orbitals. As a result of these transitions, characteristic energies are released by the atom in the form of X-ray photons.

Formation A mappable body of rock strata that consists of a certain lithologic type or combination of types. The fundamental lithostratigraphic unit.

Fractionation Crystallization from a magma in which the early formed crystals are prevented from equilibrating with the parent liquid resulting in a series of residual liquids that have a more extreme composition than would have resulted from a continuous reaction. In petrology, the residual liquids can be extruded and have a very different elemental composition from the parent magma.

FWHM "Full width half maximum" of the peak is an expression of the extent of a function, given by the difference between the two extreme values of the independent variable (this would be the elements of interest) at which the dependent variable is equal to half of its maximum value. In EDXRF this is also called resolution, calculated as the distance in eV between left and right sides of the peak at half of its maximum height.

Gamma radiation The form of electromagnetic radiation of very short wavelengths and high energy emitted by atoms undergoing radioactive disintegration. Since they penetrate matter, gamma rays are useful in radiography analyses.

Geochemical facies Any areal geological entity that is distinguished on the basis of trace element composition. (e.g., the Antelope Creek geochemical facies of the Mule Creek source).

Geochemistry The study of the distribution and amounts of chemical elements in minerals, ores, rocks, soils, and water on the basis of the properties of their atoms and ions through time and space.

Geologic province A large region characterized by similar geologic history and development (e.g., the Basin and Range Province of western North America).

Gravel An unconsolidated natural accumulation of mainly rounded rock fragments, mainly larger than sand (diameter greater than 2 mm) and may contain boulders, cobbles, pebbles or any combination; the unconsolidated equivalent of conglomerate.

Glass A solid material usually in the condition of a super-cooled liquid, formed when a molten mass of inorganic solids cools rapidly, without crystallizing. Natural glass is usually called obsidian, a rhyolite glass, but can also form from intermediate to mafic lavas under restricted conditions.

Gravimetric analysis Generic term for the classical methods of quantitative chemical analysis based on the measurement of weight.

Half-life For radioactive elements, the time required for one-half of the element to decay.

Hand specimen A piece of rock of a convenient size for megascopic study and for use in a reference collection.

Hydration A chemical reaction in which free water reacts with a solid to form hydrous materials, i.e., obsidian to perlite. In a glass, such as obsidian, a newly exposed surface hydration theoretically occurs at a regular and measurable rate, although environmental variables can intervene and cause the rate to fluctuate.

Hypabyssal A general term applied to minor intrusions such as dikes or sills and the rocks that compose them, such as felsite. These rocks often exhibit a finer grained fabric than the host lava due to more rapid cooling.

Igneous Said of a rock formed from the solidification of magma.

Ignimbrite Rock formed by the widespread deposition and consolidation of ash flows. The term includes welded tuffs and nonwelded, but recrystallized ash flows. In the western hemisphere ignimbrites tend to be defined as welded tuffs, while in European classification they tend to also include nonwelded tuffs.

Incompatible elements An element (usually trace elements) that does not substitute for major elements in crystal lattices of minerals and is instead concentrated in the melt during evolution of the magma chamber.

Infinite thickness The thickness, beyond which enough incident X-rays escape such that it is no longer possible to predict a calculated elemental quantity from a given. This is particularly an issue in XRF at high energies (i.e., 50 keV) with small samples such as obsidian debitage.

Intensity The number of X-rays counted by the detector at a given energy level or range of energy levels.

Interbedded Beds lying between or alternating with others of different character, especially contemporaneous lava flows *interbedded* between other sediments.

Interior drainage basin See closed basin.

Ion An atom or group of chemically bound atoms that have either a positive or negative charge.

Intrusion The process of emplacement of magma in or below preexisting rock.

Isomer One of two substances whose composition is identical except that the atoms in their molecules are arranged in different forms.

Isotope One of two or more species of the same chemical element – having the same number of protons in the nucleus, but with a different number of neutrons. See radioisotope. Isotopes may be of natural or artificial origin, many useful in geoarchaeological studies.

Isotropic Any medium whose properties are the same in all directions. In stone fracture mechanics, obsidian is isotropic in that force applied to any surface will travel at equal speed and force in all directions (i.e., Hertzian force).

Jasper A variety of chert containing clay and iron oxide "impurities" within the quartz crystals that impart various colors particularly red and yellow. In the archaeological vernacular used to denote any chert or chalcedony colored red or yellow.

Jet A dense relatively soft black lignite sometimes mistaken for opaque obsidian.

K-feldspar Potassium feldspar.

K spectra K spectra (i.e., K) arise following the transference of electrons to K shell vacancies. K spectra are relatively simple and consist of two doublets (K and K) with an extra line occurring for higher atomic numbers. In nondestructive EDXRF and WXRF analysis K spectra are most frequently analyzed in the mid-Z X-ray region due to very high energies, and L spectra (filling L level vacancies) for the higher energy elements such as Ba (see Chap. 2).

Lamellar flow Flow of liquid in which the layers glide over each other.

Lapilli Rock fragments between 2 and 64 mm (0.08–2.5 in.) in diameter that were ejected from a volcano during an explosive eruption are called lapilli. Lapilli (singular: lapillus) means "little stones" in Italian. Lapilli may consist of many different types of tephra, including scoria, pumice, marekanites, and reticulite.

Lattice   A regularly spaced, periodically repeated three-dimensional arrangement of points in space that specify the position of ions, atoms, or molecules in crystals.
Lava   Fluid molten rock that issues from a volcano or fissure; also the same material solidified by cooling.
Lava dome   Lava domes are rounded, steep-sided mounds built by very viscous magma, usually either dacite or rhyolite. Such magmas are typically too viscous (resistant to flow) to move far from the vent before cooling and crystallizing. Domes may consist of one or more individual lava flows. Classic domes that produced obsidian can be seen at Los Vidrios, Sonora, and Government Mountain in the San Francisco Volcanic Field in northern Arizona.
Lava flow   A lateral surficial outpouring of lava from a vent or fissure; also the same material solidified by cooling.
Law of superposition   In any sequence of sediments or igneous rocks that has not been overturned, the lowest strata will be older than the highest strata, and each bed is younger than the one beneath. First stated by Steno in 1669.
Layer   A bed or stratum of rock or sediment.
Lens   A body of rock that is thicker in the middle than at the ends; Adj. lenticular.
Liquidus   The locus of points in a temperature composition diagram representing the maximum saturation of a solid component or phase in the liquid phase.
Lithic   In geology a sediment or pyroclastic deposit containing abundant fragments of previously formed rocks. In geology and archaeology, pertaining to or produced from stone − lithic artifacts.
Lithostratigraphic unit   A body of rock that consists chiefly of a certain litholgic type or combination of types. It has a geographic name from the type area combined with a descriptive term, i.e., Coconino Sandstone.
L spectra   See K spectra and Chap. 2.

Magma   Naturally occurring molten rock material generated within the earth from which igneous rocks are derived, comprised of liquid silicate melt, suspended crystalline solids, and gas bubbles.
Magma chamber   A reservoir of magma in the shallow portion of the lithosphere (i.e., crust) from which volcanic material is derived.
Magma mixing   The mixing of two magmas to form a hybrid. Some rhyolite magmas that produced obsidian in the Southwest seem to be, in part, formed or derived from some magma mixing, such as Red Hill, New Mexico, and the San Francisco Peaks glass.
Magmatic differentiation   The process of developing more than one igneous rock type from a common in situ magma chamber, i.e., rhyolite to andesite.
Mantle   The zone of the earth below the crust divided into an upper and lower portion.
Marekanite or marekenite   Derived from eroded nodules of unhydrated obsidian that are part of the sediment load in the Marekanka River into the Okhotsk Sea of Eastern Russia. It applies specifically to nodules of obsidian, generally Tertiary in age in western North America, and called Apache Tears in the vernacular.

Mass absorption effects Mass absorption effects result from fluorescence radiation being absorbed by coexisting elements (causing reduced intensity), or enhancement of fluorescence radiation due to secondary radiation from itself or coexisting elements (causing increased intensity).

Mass spectrometer A instrument for separating atoms or ions of different mass and that can measure the exact mass of single atoms. Abbreviated as MS. Is often used in tandem with other instruments, i.e., accelerator mass spectrometer and ICP-MS

Matrix (1) The major constituents of a material in XRF analysis (see trace elements). (2) The groundmass of an igneous rock.

Matrix effect The constituent parts and elements in a substance that serve to effect the photons in highly complex ways. The complexities are collectively known as *matrix effects* which can be subdivided into overlap effects and mass absorption effects. The matrix effects on element i are the combination of mass absorption effects and overlap effects exerted on element i, by all coexisting elements j.

Medium grained An igneous rock in which the crystals have an average diameter in the range of 1–5 mm.

Melt A liquid, fused rock.

Mode The actual mineral composition of a rock expressed in weight or volume percent.

Molecule The smallest unit in which a substance can be divided and still retain it's properties and composed of one or more atoms.

Multi-channel analyzer Sorts detector output pulses according to energy level and counts the number of pulses accumulated at each level; from this information a spectrum (or pulse height analysis) is generated. From these data, the computer software can calculate elemental concentrations based on the instrument calibrations.

Neutron activation analysis An instrumental method of chemical analysis based on nuclear activation reactions – the atoms or stable isotopes or elements in a sample are identified by activating the sample by neutron bombardment and then identifying and measuring the characteristic radiation each activated element emits in relation to the analysis of standards (see Chaps. 7 and 8 here).

Nodule A small rounded lump or mass or a mineral or mineral aggregate, contrasting in composition to the surrounding rock matrix as in a nodule of obsidian in perlite.

Obsidian Obsidian is a natural rhyolite glass, a super-cooled liquid that is liquid in all its properties except in its ability to flow easily. As a glass, its atomic structure, by definition, is entirely disordered. Because of this property it has no preferred direction of fracture and is entirely isotropic, at least when entirely aphyric. This property endows obsidian with its excellent flaking properties and extremely sharp edges when fractured. Compared with window glass, obsidian is rich in iron and magnesium; tiny (<0.005 mm) crystals of iron oxide within the glass cause its dark color. Most sources of obsidian produce at least some red or mahogany colored nodules, some sources particularly in the southern Cascade Range of California and Oregon, are dominated by red obsidian. While it is certain that this coloring is produced by oxidation of the iron in the glass, it is not clear how the process transforms some portions of the flow and not others. There is no compositional difference between the red and black portions of a single nodule.

Opaque   Said of a mineral that is impervious to transmitted light. Cf. translucent; transparent.

Outcrop   That portion of a geologic formation that appears on the surface; *v.* To appear exposed on the surface.

Oxidation   Once referred to a chemical reaction in which a substance combined with oxygen, but now refers to any reaction in which a substance loses electrons.

Parental magma   The magma from which a particular igneous rock solidified.

Parent element   The radioactive element from which a daughter element is produced by radioactive decay.

Particle induced X-ray emission (PIXE)   A physical method of chemical analysis based on proton irradiation of a material. X-rays are re-emitted as a consequence of the irradiation and measured similar to EDXRF.

Particle size   The average diameter or volume of the particles in a sediment or rock.

Patina   A visible colored layer produced on the surface of a rock by weathering processes.

Peak(s)   Channel in the spectrum containing the highest number of counts within a distribution of counts. The height and overall area of peaks within a spectrum yield quantitative information about the element(s) present within a sample.

Peak count   The sum of all counts that fall inside the ROI. See ROI.

Pebble   A rock fragment generally rounded by abrasion, larger than a granule and smaller than a cobble with a diameter in the range of 4–64 mm. The vast majority of marekanites derived from Tertiary sources in North American obsidian sources fall within this range.

Peralkaline   A division of igneous rocks in which the molecular proportion of alumina (AlO_2) is less than that of sodium and potassium oxides combined. Peralkaline obsidians in the Southwest include many in the basin and range region of Chihuahua, including Antelope Wells on the New Mexico/Chihuahua border. These peralkaline obsidians are characterized by high iron and zirconium relative to rubidium and strontium, and are frequently dark green in color from the high iron content.

Peraluminous   A division of igneous rocks in which the molecular proportion of alumina (AlO_2) exceeds that of sodium and potassium oxides combined. Many of the obsidian sources north of the US/Mexican border in the U.S. Southwest fall into this category (i.e., Antelope Wells/El Berrendo and Los Sitios del Agua).

Perlite   A volcanic glass with a composition of rhyolite (obsidian) that has a higher water content than obsidian. Perlite is the eventual crystalline end for obsidian when it has completely hydrated.

Perlitic structure A feature of volcanic rocks, particularly glassy rhyolites, that have cracked due to contraction during cooling forming small concentric pearl-like spheroidal structures. This occurs at megascopic and microscopic levels. Marekanites are often found as embedded remnants within perlite.

Petrography The branch of geology, specifically petrology, dealing with the description and systematic classification of rocks by means of microscopic examination of thin sections.

Petrology The study of rocks that form the earth's crust.

Phenocryst One of the relatively large megascopically visible crystals of the earliest generation in a porphyritic volcanic rock. In obsidian, the visible phenocrysts are most often sanidine feldspar.

Pitchstone Obsidian with a higher proportion of water and as a result is generally vitrophyric and crystalline and a poor media for tool production. Geologists often map pitchstone as obsidian, but in the Southwest artifact quality obsidian rarely occurs in association with pitchstone. Tank Mountains, Vulture, and Sauceda Mountains source areas exhibit pitchstone, but no artifact quality marekanites have been found in direct association, while the Los Sitios del Agua source in northern Sonora exhibits aphyric marekanites in a perlitic matrix that resembles a pitchstone. The East Grants Ridge source in the Mount Taylor Volcanic Field appears to be a pitchstone in hand sample, but is still an adequate media for tool production.

PIXE See particle induced X-ray emission.

Plagioclase A group of triclinic feldspars of the general formula $(Na, Ca)Al(Si, Al)Si_2O_8$. Plagioclase is one of the most common rock forming minerals, particularly in intermediate and mafic volcanic rocks. It does occur in high silica rhyolites, but less commonly than sanidine.

Plate boundary A zone of seismic and tectonic activity along the edges of lithospheric plates. The Tertiary activity along and underneath the plate boundary between the Pacific and North American plate was instrumental in creating the volcanism the produced most of the Tertiary obsidian sources in the Southwest.

Pleistocene The epoch of the Quaternary Period after the Pliocene of the Tertiary and before the Holocene beginning about 2–3 million years ago and lasting until the Holocene about 8,000 years ago. The obsidian sources along the southern scarp of the Colorado Plateau and associated with the formation of the Valles Caldera in northern New Mexico were formed during the Pleistocene. Most of the secondary deposits of Tertiary sources are in Pleistocene formations, such as the 111 Ranch Formation in the Safford and San Simon Valleys of southeastern Arizona that contain pebbles of Cow Canyon and Mule Creek Tertiary obsidians.

Plinian eruption An eruptive event in which a steady, turbulent stream of fragmented magma and magmatic gas is released at high velocity creating large volumes of pyroclastics and high eruption columns. Named for Pliny the Younger who was a surviving eye witness to the A.D. 79 eruption of the stratovolcano Vesuvius in southern Italy. Some of the obsidian in the Southwest was formed by smaller Plinian type eruptions where the pyroclasts quenched during the eruption and are preserved in the ash or ignimbrites; i.e., Los Vidrios, Mule Mountain (Mule Creek), some of Sand Tanks (Shackley 2005).

Pliocene   The epoch of the late Tertiary after the Miocene and before the Pleistocene. The obsidian in the Mount Taylor Volcanic Field (Grants Ridge and Horace Mesa) were formed at the boundary of the Pliocene and Pleistocene, in part due to tectonic activity associated with the uplift of the Colorado Plateau (Shackley 2005).

Plutonic rock   A rock formed at considerable depth by slowly cooling magma. It is characterized by medium to coarse grained texture – granitoid. Granite is the plutonic form of rhyolite.

Porphyritic   A type of inequigranular texture in which there is a distinctly bimodal population of grain sizes. The large grains are phenocrysts surrounded by a finer grained matrix or groundmass. Porphyritic rocks occur in fairly small relatively shallow intrusive rock bodies (i.e., hypabyssal rocks) that have experienced extended period of relatively uniform crystallization. Many rhyolites are porphyritic due to this process even though they have been extruded. Obsidian that is porphyritic are called vitrophyric.

Potassium feldspar   An alkali feldspar containing the molecule ($KalSi_3O_8$), such as orthoclase, and sanidine. K-feldspar in the vernacular and common in silicic rocks such as rhyolite.

p.p.m   The abbreviation for parts per million a form expressing the concentration of substances (i.e., elements) highly diluted in a material. It is equivalent to 1 g per 1 metric ton. Often further abbreviated as "ppm."

Provenance   As used here, the geographical source of a material, either primary or secondary.

Precision   The ability of an instrument to obtain consistent results when performing multiple measurements on the same sample (see also repeatability).

Pulse   Analog output waveform produced by the XRF detector, electronics and amplifier. Each pulse is proportional in magnitude to the energy of a detected X-ray photon (see multi-channel analyzer).

Pyroclastic flow   A pyroclastic flow is a ground-hugging avalanche of hot ash, pumice, rock fragments, and volcanic gas that rushes down the side of a volcano as fast as 100 km/h or more. The temperature within a pyroclastic flow may be greater than 500°C, sufficient to burn and carbonize wood. Once deposited, the ash, pumice, and rock fragments may deform (flatten) and weld together because of the intense heat and the weight of the overlying material. A number of pyroclastic flows produced obsidian in the Southwest including Los Vidrios, Sonora.

Qualitative analysis   The study of the *identity* of the components of a substance; as opposed to quantitative analysis that serves to yield the proportions or absolute quantities of components.

Quantitative analysis   The study of the exact relative amounts of each component of a substance. In obsidian analysis the elements and compounds are reported in parts per million or percent by weight.

Quarry *Open* workings for the extraction of stone. As opposed to an adit or mine.

Quaternary The second period of the Cenozoic Era, following the Tertiary that began about 2–3 million years ago and consisting of two unequal epochs;; the Pleistocene followed by the Holocene that began about 8,000 years ago.

Quenching Rapid or essentially instantaneous cooling of a material such that crystallization is hindered or eliminated. Obsidian is a quenched rhyolite such that it is completely disordered with no crystalline form.

Radioactive daughter A nuclide produced by the radioactive decay of another nuclide known as the radioactive parent.

Repeatability The ability of an instrument to obtain consistent results when performing multiple measurements on the same sample – the same as precision.

Reproducibility The ability of an instrument to obtain consistent measurement results when measuring the same sample at different times and /or with different operators and/or using different instruments of the same type (see Chap. 2).

Region of interest (ROI) The region of interest is used to calculate peak counts. The ROI depends on the element as well as the detector type used to conduct the test. For example, with a Cu sample on a Peltier cooled detector system like the ones in this volume, the ROI would be defined as ± 5 channels or ± 100 eV around the tallest channel in the very first spectrum analyzed during the test. To allow for advanced data acquisition, the ROI width can be modified manually prior to the start of the test.

Rhyolite A group of silicic extrusive igneous rocks, often porphyritic and commonly exhibiting flow banding, with phenocrysts of quartz, and alkali feldspars (i.e., sanidine) in a glassy cryptocrystalline groundmass. Rhyolite is commonly light colored, but can be as dark as basalt, and is sometimes confused with basalt in archaeology. The Presley Wash rhyolites in the Mount Floyd Volcanic Field in northern Arizona are typical of dark glassy rhyolites. The extrusive equivalent of granite. Quenching of rhyolite lava can create obsidian.

Rhyolitic magma Erupted at temperatures of 750–1,000°C, rhyolitic magma is comprised mainly of SiO_2 (75 wt.%), Al_2O_3 (13%), and the alkalis Na_2O and K_2O in about equal amounts (3–5%). Due, in part, to the extreme viscosity of rhyolite magma, little is actually erupted, but remains in the crust as plutons, and is eventually uplifted or otherwise exposed, such as the Sierra Nevada. Rhyolitic magma in what is now the Southwest is often produced by reprocessing (remelting) older crust. In western Arizona, Precambrian granites became the raw material, through reprocessing during the Tertiary, for rhyolite and when quenched, obsidian, such as Sand Tanks and Tank Mountains, both of which are surrounded by older Precambrian granite basements.

Rock An aggregate of one or more minerals, e.g., quartz, granite, shale; or disordered mineral matter, e.g., obsidian, or solid organic matter, e.g., coal.

Röntgen, Wilhelm Wilhelm Conrad Röntgen (1845–1923) who was a Professor at Würzburg University in Germany discovered X-rays in 1895.

Sanidine   A high temperature mineral of the alkali feldspar group, $KalSi_3O_8$. It is a disordered monoclinic feldspar occurring in clear, glassy crystals in unaltered silicic volcanic rocks such as obsidian, rhyolite, and dacite.

Scintillation counter   An instrument that measures ionizing radiation by counting the individual scintillation of a substance. The primary counting method in X-ray fluorescence spectrometry.

Sediment load   The solid material transported by a stream system, expressed as the dry weight of all sediment that passes a specified point in a given period of time.

Silicic   Silica rich igneous rock or magma. The amount of silica is generally said to constitute 65% or more or two thirds of the composition of the rock. These rocks generally contain free silica in the form of quartz; e.g., obsidian, rhyolite, granite, dacite. Formerly called acid igneous rocks.

Solidus   The temperature below which a magma is completely crystallized, or in the case of obsidian the point at which the glass quenches.

Spectroscopy   The study of the properties of electromagnetic radiation with a spectroscope; e.g., X-ray fluorescence spectroscopy (also spectrometry).

Spectrum   A frequency of occurrence histogram displaying the number of detected X-rays (counts) along a Y vertical axis and their respective energy levels (in keV) along its X horizontal axis; used to make qualitative and quantitative determinations about sample materials (see Figs. 2.2 and 2.7).

Spherulite   A rounded mass, usually feldspar, of acicular crystals radiating from a central point. Sometimes forming in obsidian, it is common in the Cerro Toledo Rhyolite glasses (spherulitic obsidian) at Valles Caldera in northern New Mexico, particularly those erupted from the Rabbit Mountain center, a dome on the southeastern rim of the caldera.

Standard   A sample material of known thickness and/or composition used to calibrate XRF units for specific applications (see calibration). Many standards are internationally recognized and used by the laboratories in this volume.

Stratigraphy   The arrangement of rock strata as to chronological sequence.

Stratovolcano   A volcano, usually intermediate lava, constructed of alternating strata of lava and pyroclastic deposits, usually with abundant dikes and sills. Also called a composite volcano or cone.

Stratum   A layer of sedimentary rock visually separable from layers above and below; a bed; plural strata.

Surface analysis   The general term for a number of chemical analytic methods that analyze the surface of a substance. Includes X-ray fluorescence spectrometry, both energy-dispersive and wavelength dispersive.

Tektite   A silicate glass of nonvolcanic origin, generally formed by extra-terrestrial impacts on terrestrial rocks. They often resemble marekanites and can have very similar composition with much less water (average 0.005%), but more often have a composition more similar to shale.

Tephra   Tephra is a general term for fragments of volcanic rock and lava regardless of size that are blasted into the air by explosions or carried upward by hot gases in eruption columns or lava fountains. Tephra includes large dense blocks and bombs, and small light rock debris such as scoria, pumice, reticulite, and ash, and rhyolite lava fragments can be quenched into obsidian under the right conditions. As tephra falls to the ground with increasing distance from a volcano, the average size of the individual rock particles becomes smaller and thickness of the resulting deposit becomes thinner. Small tephra stays aloft in the eruption cloud for longer periods of time, which allows wind to blow tiny particles farther from an erupting volcano.

Ternary system   A system of three components often displayed in an equilateral triangular graph. Early qualitative obsidian analyses used ternary graphs to display the relative proportions of elemental components, i.e., Rb, Sr, Zr.

Tertiary   The first period of the Cenozoic Era after the Cretaceous of the Mesozoic Era and before the Quaternary. It is generally agreed that it covers the time span from 65 million to 2 million years ago, and is applied to the corresponding system of rocks. Divided into unequal epochs; Paleocene, Eocene, Oligocene, Miocene, and Pliocene. The earliest dated, still artifact quality, obsidian in the Southwest is Antelope Creek and Mule Mountains of the Mule Creek caldera that erupted during the Miocene at about 17 million years ago. Most of the Basin and Range obsidian in the Southwest dates to the middle to late Tertiary.

Total counts   The sum of all counts in the spectrum between 400 and 41,960 eV, the effective range of acquisition for EDXRF ($Z = 11$–92).

Trace element   A chemical element which while a component of a material is not essential to the composition of the material and occurs in very low concentrations usually below 0.01% (100 p.p.m.), but often up to 0.1% (1,000 p.p.m.) in the XRF vernacular. The concentration of trace elements are often unique signatures of a material, such as many of the incompatible elements in obsidian.

Transition elements   The elements located between Group IIA and group IIIA in the periodic table.

Tuff   A general term for all consolidated pyroclastic rocks. Not to be confused with the chemical sedimentary rock tufa.

Tufflava   An extrusive rock containing pyroclastic and lava-flow characteristics and is considered to be an intermediate form between a lava flow and a welded tuff form of ignimbrite. The Mule Mountain/Mule Creek obsidian and some of the Los Vidrios obsidian were probably formed in tufflava eruptions.

Type locality   The place from which a geologic feature, such as a specific igneous rock type, was first recognized and described; e.g., marekanite from the Marekanka River obsidian nodules in Eastern Siberia.

Unconformity   A break or gap in the stratigraphic record such as an abnormal break in the sequence of sedimentary deposition; i.e., older strata found above younger strata.

Unconsolidated material   Sedimentary material that is loosely associated in a stratum or on the surface, and whose particles are not cemented.

Valence   The property of atoms to combine chemically with only one definite number of other atoms.

Vesicular   Characterized by containing vesicles as in vesicular basalt.

Vitreous   Having the luster and appearance of glass.

Vitric tuff   An indurated deposit of volcanic ash composed mainly of fragments of glass ejected during eruption.

Vitrophyre   Any porphyritic volcanic rock having a glassy groundmass; generally restricted to rhyolite and dacite. Generally applied to non-aphyric obsidian.

Vitrophyric   Of or pertaining to vitrophyre. In the vernacular pertaining to perlitic and non-aphyric obsidian similar to pitchstone.

Volcanic ash   Fine pyroclastic ejecta under 2 mm in diameter. Usually refers to unconsolidated material but sometimes used to refer to tuff.

Volcanic breccia   Angular pyroclastic volcanic fragments that are larger than 64 mm in diameter.

Volcanic dome   A rounded protrusion of usually silicic lava that is squeezed out from a volcano forming a dome shape or bulbous mass above and around the vent.

Volcanic glass   Natural glass produced by the quenching of lava before crystallization can occur. Most common in silicic lavas where silicon and aluminum oxides are relatively high. Includes both obsidian and intermediate and mafic glasses.

Vug   A small cavity in a vein or rock usually lined with crystals of different chemical composition. Often chalcedony is deposited through precipitation in rhyolite and andesite lavas in the Southwest, in some cases offering artifact quality raw material for stone tool production.

Wavelength   The measured distance between successive wave crests, valleys, or other equivalent points in a wave of electromagnetic radiation or sound. The accepted symbol is ".".

Weathering   The mechanical or chemical destruction of a substance through interaction with the atmosphere with little or no transport of the substance.

Welded tuff   A glass rich pyroclastic rock that has been subjected to subsequent burial by other pyroclastic rock and through heat and compaction has welded the glass shards in the matrix. Often appears banded or streaked, and is extremely dense and hard; also called ignimbrite.

whole-rock analysis   A procedure whereby the entire rock is analyzed rather than a constituent. X-ray fluorescence spectrometry is a whole-rock analysis.

Xenocryst   A phenocryst in volcanic rock that resembles a native mineral, but is actually foreign to the environment in which it occurs.

Xenolith   A foreign intrusion into igneous rock larger than a mineral.

X-ray   Radiation of extremely short wavelength produced by the bombardment of a substance by a stream of electrons moving at high velocity.

X-ray diffraction analysis   A technique for identifying and analyzing the structure of crystalline materials by the three-dimensional periodic array of atoms in a crystal that has periodic repeat distances (lattice dimensions) of the same order of magnitude as the wavelength of the X-rays. X-ray diffraction has been used in the analysis of minerals in ceramics with some success. Since there is really no mineral lattice dimensions in a glass, it is not useful in the analysis of obsidian.

X-ray filters   See filter(s).

X-ray fluorescence analysis (see Chap. 2)   A potentially nondestructive method of chemical analysis (XRF) based on the use of high energy X-rays. An intensive beam of X-rays is used to irradiate a sample causing it to re-emit (fluoresce) radiation the wavelength (WXRF) or energy (EDXRF) of which is used to determine the elemental composition of the substance. XRF is a mass analysis method. All material that is subjected to irradiation is included in the analysis. Therefore, heterogeneous substances can issue spurious results if the area irradiated is too small to represent the modal character of the substance. Aphyric obsidian is an excellent media for EDXRF analysis since it is a homogeneous non-crystalline substance, and many of the incompatible elements that discriminate magma sources are well measured by EDXRF.

X-ray tube   An evacuated enclosure containing an anode (or filament) and a quantity of target material. When high voltage is applied to the anode, electrons collide with the target material, inducing it to fluoresce X-ray photons.

Z   Equals atomic number (i.e., $Z = 37 = Rb$).

References

Bates, R.L. and Jackson, J.A., (Eds.), (1984), *Dictionary of geological terms.* Third Edition. Prepared by the American Geological Institute. New York: Anchor Books/Doubleday.

Goffer, Z., (1996), *Elsevier's dictionary of archaeological materials and archaeometry*: In English, with translations of terms in German, Spanish, French, Italian, and Portuguese. Amsterdam: Elsevier.

Jenkins, R., (1999), *X-ray fluorescence spectrometry.* Second Edition. New York: Wiley-Interscience/Wiley.

Shackley, M.S., (2005), *Obsidian: geology and archaeology in the North American Southwest.* Tucson: University of Arizona Press.

Sigurdsson, H., (Ed.), (2000), *Encyclopedia of Volcanoes.* San Diego: Academic.

Thorpe, R. and Brown, G., (1985), *The field description of igneous rocks.* Chichester: Wiley.

Index